Name:

Phone:

Email:

Class:

Room:

These are my class notes, and I really need them to study. If you find this book, please contact me or leave the book in the classroom where I can find it. Thank you!

Organic Chemistry 2
Lecture Guide

Rhett C. Smith, Ph.D.

In cooperation with Iron Foundation Publications

Before We Begin:
A Brief Review of Key Concepts and Reactions from Organic Chemistry 1

Suggested Reading:

Suggested Problems:

Rules of the Game

1. Opposite Charges Attract

Attractions = better stability

2. Like Charges Repel

Repulsion = less stability

Notes:

Hybridization and Molecular Orbitals

Two Orbitals can combine ...

<u>**recommended:**</u>
You should be able to sketch an orbital resulting from the addition of any two orbitals given.

... by adding ...

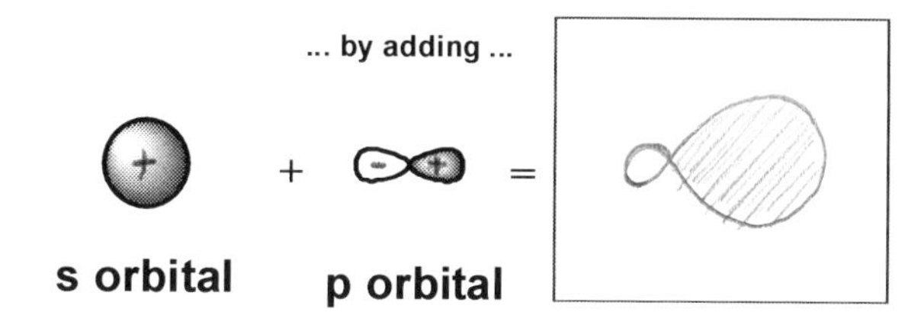

... or by subtracting ...

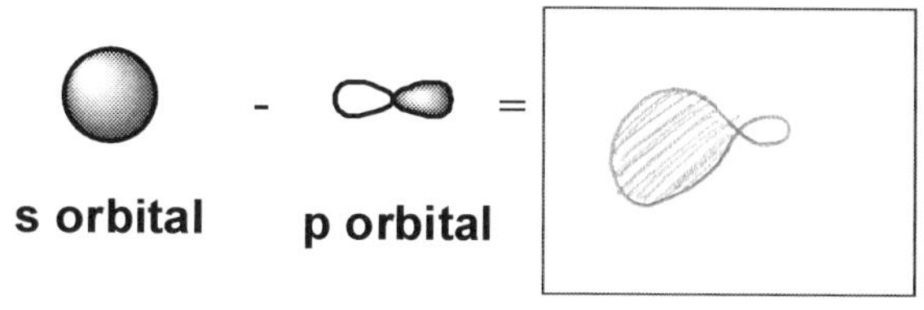

... to form two new orbitals.

Adding or "hybridizing" one s orbital with one p orbital gives two **sp hybrid orbitals**.

(not to scale)

Notes:

- Each orbital can hold up to 2 e^-.
- Combining 2 atomic orbitals gives 2 hybrid orbitals.

The sp hybridized atom

Drawing both sp hybrid orbitals on the same atom:

This is an **sp hybridized** carbon atom. The sp orbitals are pointed directionally opposite from one another, in what is known as a **linear geometry**.
Since carbon has three p orbitals and we only used one to make the hybrids, two p orbital are still left unhybridized on carbon.

The remaining p orbitals lie perpendicular to the line of the sp hybrid orbitals.

Using the dash-wedge notation, we can denote the geometry and show the p orbital that remains unhybridized.

Notes:

- Hybrid orbitals are only needed for:
  1. σ bonds. (i.e. bonds w/ other atoms / not π bonds)
  2. holding lone pairs.

The sp^2 hybridized atom

Three **Orbitals can add together ...**

+ + = + +

s orbital

... to form three new orbitals.

Adding or "hybridizing" one s orbital with two p orbital gives three **sp^2 hybrid orbitals**.

Notes:

The sp^2 hybridized atom

Drawing all three sp^2 hybrid orbitals on the same atom:

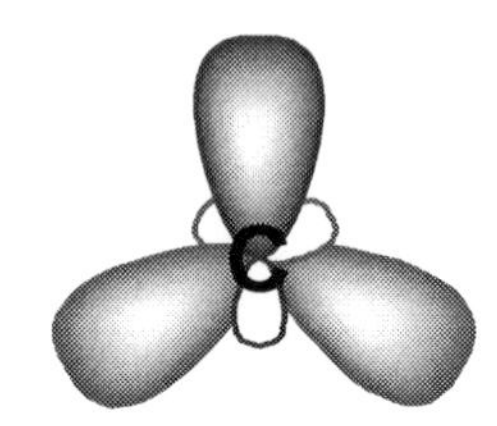

This is an **sp^2 hybridized** carbon atom. The sp^2 orbitals are in a plane pointed directionally______° from one another, towards the corners of an equilateral triangle, in what is known as a ________________________________geometry.

Since carbon has three p orbitals and we only used two to make the hybrids, a p orbital is still left unhybridized on carbon.

It lies perpendicular to the trigonal plane of the three sp^2 hybrid orbitals.

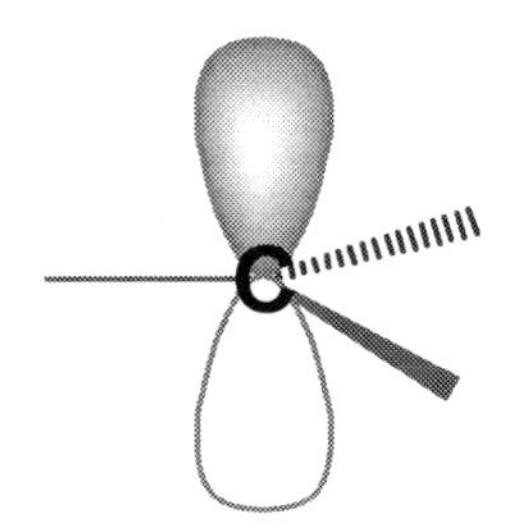

Using the dash-wedge notation, we can denote the geometry and show the p orbital that remains unhybridized.

Hybrid orbitals are involved in forming, or may be filled by a lone pair of electrons.

Notes:

The sp^3 hybridized atom

Four **Orbitals can add together ...**

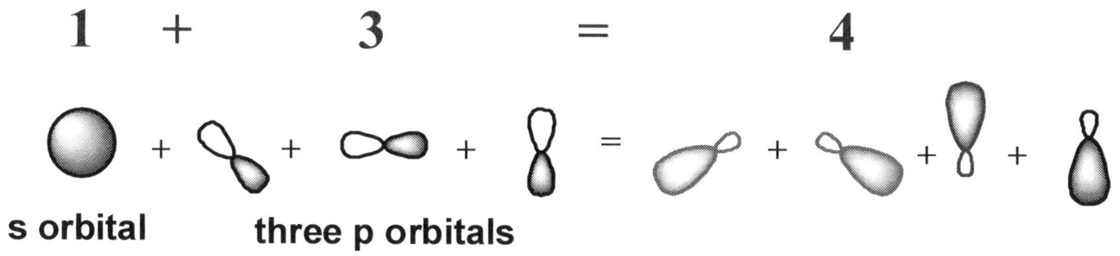

1 + 3 = 4

+ + + = + + +

s orbital **three p orbitals**

... to form four new orbitals.

Adding or "hybridizing" one s orbital with three p orbital gives four **sp^3 hybrid orbitals**.

Notes:

The sp^3 hybridized atom

Drawing all four sp^3 hybrid orbitals on the same atom:

This is an **sp^3 hybridized** carbon atom. The sp^3 orbitals are pointed directionally ______ ° from one another, towards the corners of a tetrahedron, in a ________________ geometry.

Since carbon has three p orbitals and we used all three to make the hybrids, all of the valence orbitals are now sp^3 orbitals, and consequently these are the only ones used in bonding.

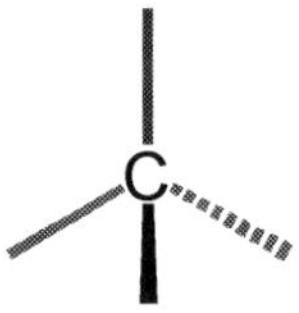

Using the dash-wedge notation, we can denote the geometry and show the p orbital that remains unhybridized.

****

Hybrid orbitals are involved in forming [ ]

, or may be filled by [ ]

Notes:

Sigma (σ) and Pi (π) Bonds

A sigma (σ) bond is one in which the electrons making the bond are between the two nuclei joined by the bond:

A pi (π) bond is one in which the electrons lie above and below the line between the two nuclei which are joined by the bond:

In the vast majority of example we will study, the **hybrid orbitals** will form sigma bonds or will be filled by a lone pair of electrons, and any pi bonds will be made by p orbitals not hybridized into sp, sp^2, or sp^3 orbitals…

Notes:

Sigma (σ) and Pi (π) Bonds

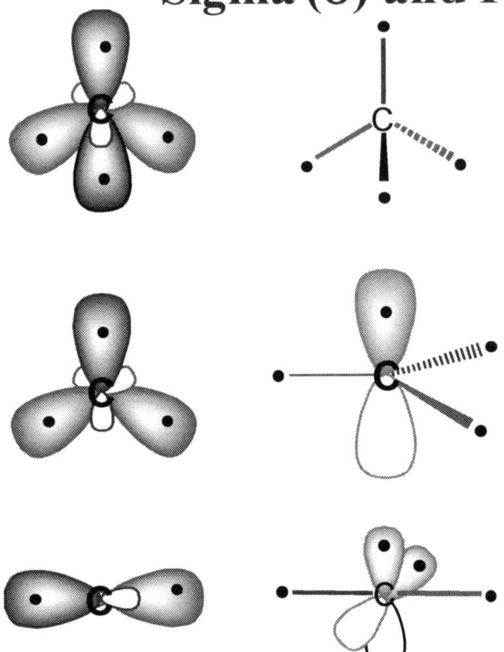

Here are the pictures we saw for hybrid orbital sets on carbon (left) and the geometry of hybrid orbitals along with leftover p orbitals (shown in grey) on the right. Recall (look at a periodic table) that a carbon atom has four valence electrons, so I have placed one electron in each valence orbital. That means that another atom must supply one electron to form a bond with any of these orbitals ...

We will consider how each type of hybridized carbon shown will form sigma bonds using the hybrid orbitals and pi bond with leftover p orbitals

Notes:

Sigma (σ) and Pi (π) Bonds

We can bind two sp^3 hybridized carbons together and make sigma bonds to hydrogen atoms with the remaining sp^3 orbitals

This is ETHANE. There are no more valence orbitals, so we can make no more bonds.

Notes:

Sigma (σ) and Pi (π) Bonds

After using all the sp^2 hybrids to make sigma bonds, we have a pi orbital left over on each carbon, each with one electron. These two orbitals are close together and have the right number of electrons (two) to form a bond. Not surprisingly, this is what happens, and this is an example of a pi bond!

Here is a cartoon of what the pi bond looks like. Note that the electrons are above and below the carbon-carbon line.
This molecule is ETHENE.

Notes:

Sigma (σ) and Pi (π) Bonds

After using all the sp hybrids to make sigma bonds, we have two pi orbitals left over on each carbon, each with one electron. These orbitals are close together and form two pi bonds, so overall there is a triple bond between the carbons!

Here is a cartoon of what the pi bonds looks like. Note that the electrons are above and below the carbon-carbon line.
This molecule is ETHYNE, commonly called ACETYLENE, a flammable gas used in welding.

Notes:

For our purposes:

1. All single bonds are
2. All double bonds are made of
3. All triple bonds are made of

Finally, a note on electronegativity of hybridized carbons:
EN of carbons increases in the order

The s orbital is closer to the nucleus than the p orbital. It is consequently filled with electrons before the p orbital, and we may think of this as demonstrating that an orbital that has more

" " has a greater pull on electrons.

Since sp is "half s and half p", and sp^3 is "$^1/_4$ s and ¾ p", it is logical that the sp hybrid has a greater pull for electrons

Notes:

Predicting Acidity

$$HA + H_2O \rightleftharpoons A^- + H^+$$

Any factors that favor dissociation of HA into H^+ (to form H_3O^+ in water) and A^- will enhance acidity.

Factors that affect dissociation:

***1.** Anion Stability

Notes:

Effect 1: Anion Size

Within a group (column), the size of the anion has a strong effect on acidity, because:

pK_a ~

Weakest	HF	3
	HCl	-7
	HBr	-8
Strongest	HI	-9

- ORBITAL OVERLAP
- **ANION STABILITY**

Notes:

- Atoms w/ charge = larger → more stable
- larger atoms have extra volume to carry charge after dissociation

Effect 2: Electronegativity (EN)

If the atoms to be deprotonated are in the same period (row) then they are about the same size, so the size effects are minimal. Thus, the electronegativity plays a strong role because:

	CH_4	NH_3	H_2O	HF
	$CH_3^{\ominus}$	$NH_2^{\ominus}$	$OH^{\ominus}$	$F^{\ominus}$
Weakest A.				Strongest A.
pK_a ~	______	______	______	______

Notes:

- Remember same row = same size
- same size but different Atomic #, so more ⊕ on nucleus = more attractive force on e^-

Effect 3: Hybridization

If the atom to be deprotonated is the same and only hybridization changes, the hybridization electronegativity plays a strong role because:

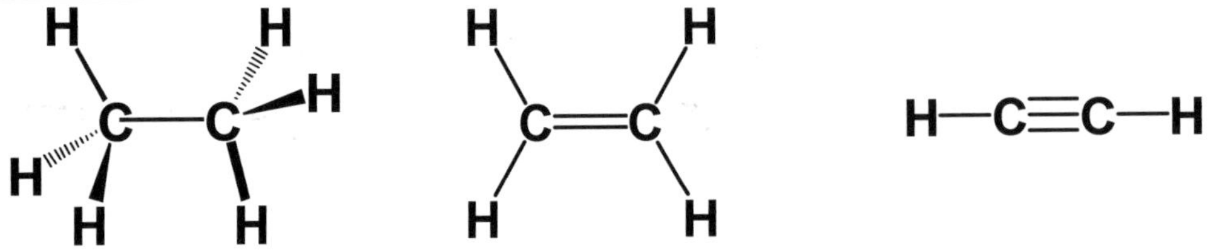

H—C≡C—H

pK_a ~ ______ ______ ______

H—C≡C

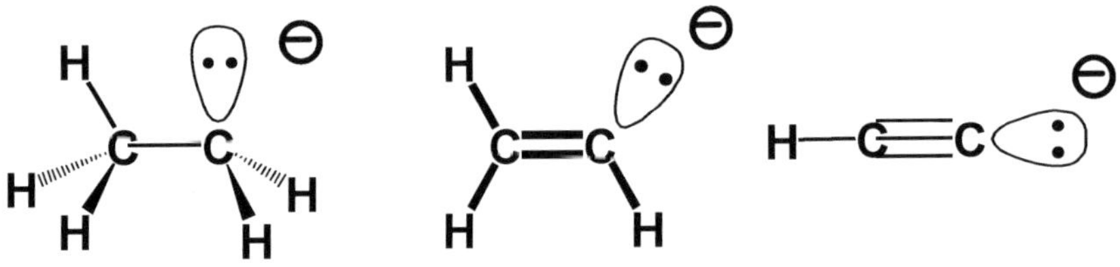

Notes:

- More "S" character = more stable anion
- sp > sp² > sp³ (hybrid stability)
- e⁻ in "s" orbital are closer to nucleus.

Effect 4: "Resonance Effects"

If the anion produced by deprotonation has more than one (good) resonance form, then:

CH_3CO_2H
$pK_a = 4.7$

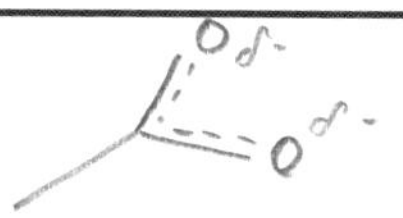

2 equal Resonance structures
Delocalization energy → spread out charge

CH_3COH
$pK_a = 15.5$

No Resonance

HCN
$pK_a = 9.1$

H - C≡N

No Resonance

Notes:

Effect 5: "Inductive Effects"

If an atom to be deprotonated has a partial positive charge INDUCED on it by nearby atoms, it is easier to deprotonate because:

This series illustrates: E.N. = More Stability
Electronegativity → e⁻ w/d effect

	CH_3COOH	ICH_2COOH	$BrCH_2COOH$	$ClCH_2COOH$	FCH_2COOH
pK_a	4.8	3.2	2.9	2.8	2.7

	2-bromopentanoic acid	3-bromopentanoic acid	4-bromopentanoic acid	5-bromopentanoic acid
pK_a	3.0	4.0	4.6	4.7

This series illustrates: the closer δ+ = greater stability

Notes:

Anion near partial ⊕ charge = more stable

(cont'd)

(Hyperconjugation)/sterics:*

Note: C is slightly more electronegative than H, so:†

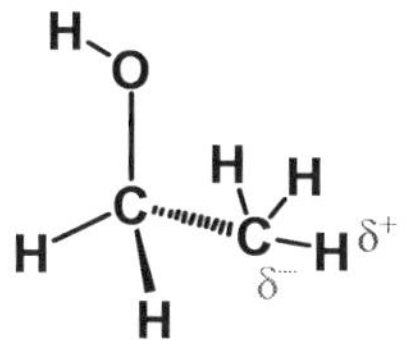

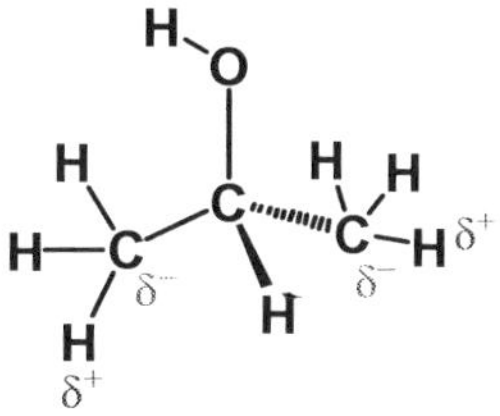

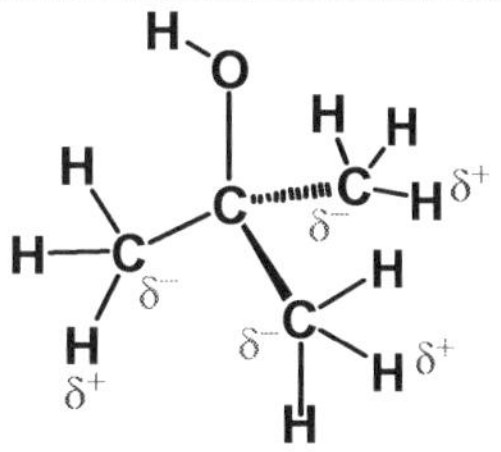

pK_a ~ ______ ______ ______

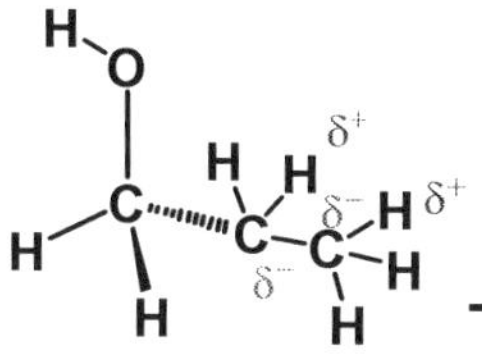

______

* More 'correct' Cartoon of hyperconjugation: 5th ed pg 102; 4th ed. Pg 144.

†Not as technically 'correct', but simple memory aid

Notes:

- Negative bonds repel e^-

Cations

Ⓐ **Observation:** more substituents = more stability

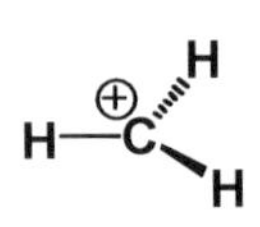

methyl **1°** **2°** **3°**

STABILITY

LESS STABLE **MORE STABLE**

Ⓑ **Explanation:**

Ⓒ **Ramification:**

REACTIVITY

$H_3C\text{-}OH < 1^o < 2^o < 3^o$

Notes:

"Markovnikov's Rule" → more subbed cation is more stable

Radicals

We've discussed the divergent reactivity of the dihalides and that only Cl_2 and Br_2 are commonly used in this radical chain reaction, but the structure of the alkane we use also plays an important role.
The key step is the homolytic cleavage of the alkane C-H bond to form the radical:

R–H $\rightarrow$ R• + H• ΔH = BDE (**bond dissociation energy**)

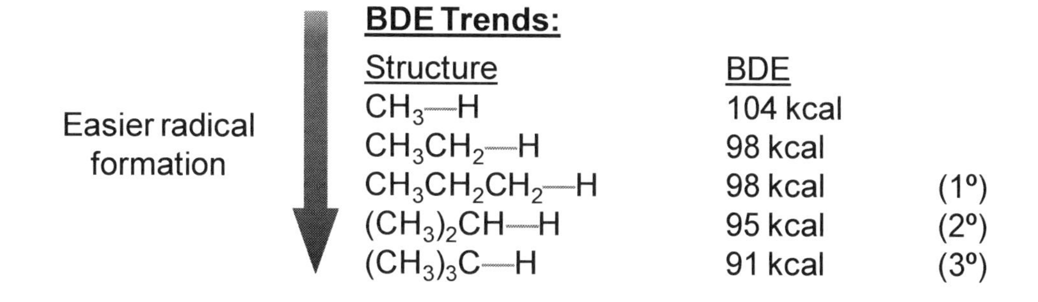

BDE Trends:

Structure	BDE	
CH_3—H	104 kcal	
CH_3CH_2—H	98 kcal	
$CH_3CH_2CH_2$—H	98 kcal	(1°)
$(CH_3)_2CH$—H	95 kcal	(2°)
$(CH_3)_3C$—H	91 kcal	(3°)

(opposite the trend we discussed for carbanions! Same trend for carbocations).

Notes:

Selectivity in radical halogenation reactions

- more subbed Radical = more stable
- Radicals are neutral
- Radicals tend to follow cation rules to complete their octet

Your step-by-step guide to alkane nomenclature

1. Find the longest chain. This is the 'parent chain'; the other stuff on the parent chain will be named as substituents. Texts and profs will draw the molecule in weird ways to try to fool you, so look out:

2. Number the carbon atoms in the 'parent chain' in the way that gives the lowest number to the substituent closest to an end of the parent chain.

3. With more than one type of substituent, name in alphabetical order.

4. If more than one of the same substituent are present on your parent chain, use di, tri, tetra, etc. prefixes to denote this (these prefixes do not count when alphabetizing, though; neither do the n-, sec-, or tert- prefixes)

5. If numbering leads to the same lowest number substituent in both directions the correct numbering gives the lowest number to the substituent that is first alphabetically.

6. If you find two different possible parent chains of the same length, you choose the one with more substituents coming off of it.

7. Substituents that are not *n*-alkyl groups will need to be named using their own nomenclature rules:

   a. Always give number 1 to the carbon attached to the parent chain
   b. The first letter of the substituent name is used for alphabetizing
   c. In the final name of the whole compound, these complicated substituent names are put in parentheses to cut down on the confusion.

Notes:

Naming Alcohols: Rules

For alcohols, use the rules for naming alkanes and cycloalkanes as a starting point, with the following adjustments:

1. replace the "e" at the end of the alkane name with "ol".

2. The alcohol is always given the lowest possible number. Note that this means that the alcohol is always given the "1" position in cycloalkanes (so, no need to add a number; we know it's always 1).

3. Place the number indicating the position of the alcohol directly before the parent chain name (which now ends in "ol").

Notes:

Naming Alkenes

Nomenclature of alkenes is related to that for alkanes, with modifications:

1. always designate the longest chain having

(A) [ ] as the parent chain.

2. Number the parent chain such that the double bond

(B) [ ]

4. change the 'ane' ending of the name you would use if it were an alkane

(C) [ ]

5. Tell where the double bond starts:

(D) [ ]

Notes:

Alkene Isomers: *cis-* or *trans-*

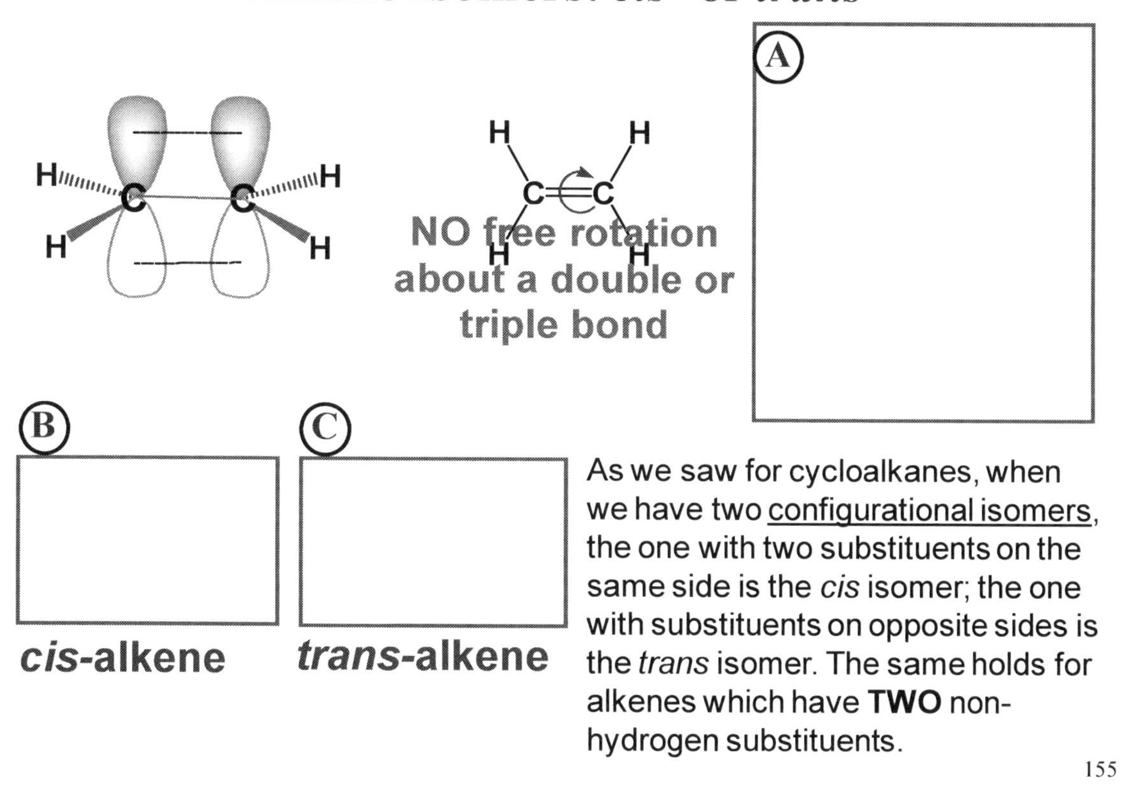

A
H
H
C
C
H
H
H
H
C
C
H
H
NO free rotation about a double or triple bond
B
C
cis-alkene
trans-alkene
As we saw for cycloalkanes, when we have two configurational isomers, the one with two substituents on the same side is the cis isomer; the one with substituents on opposite sides is the trans isomer. The same holds for alkenes which have TWO non-hydrogen substituents.
155

Notes:

- ONLY DI- subbed alkenes can be cis/trans

Easy EZ

I. The generic structures representing E- and Z- isomers should be used in deciding whether an alkene is the E- or Z- isomer:

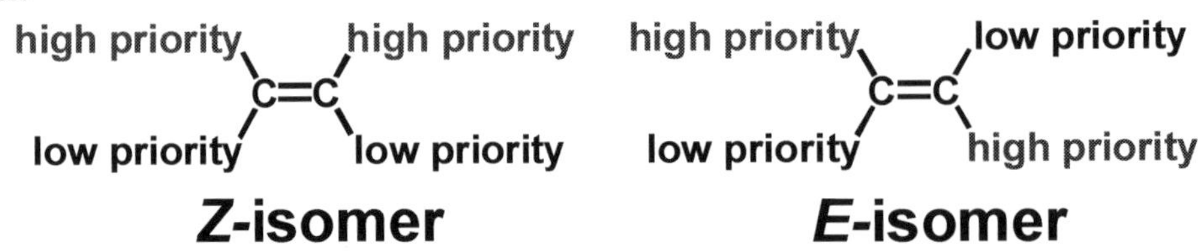

Note from the generic structures that you only need to know which substituent is high and which is low ON EACH CARBON. The best way to do this is to start by dividing the alkene into two halves:

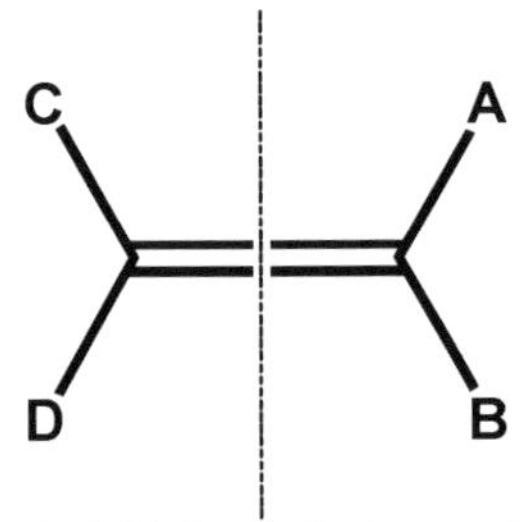

Your task is then to determine whether **A** or **B** is higher priority on the right-side carbon and whether **C** or **D** has higher priority on the left-side carbon. Once you do this, you simply match your assignments to one of the generic representations, i.e., if the higher priority substituent on the left side and right side carbons both point to the same side of the double bond (both up or both down) it is a Z- isomer, whereas if one of the high priority substituents is up and one is down (on opposite sides of the double bond) it is an E- isomer.

II. When we talk about "substituents" in the context of assigning priorities, we are talking about whatever is attached to the carbon atoms having the double bond in question; **these are NOT the same substituents that are listed in the name of the compound!** In the structure above, **A**, **B**, **C**, and **D** are the "substituents", regardless of their identity. For example, the "substituents" on the alkene below are outlined with dashed lines:

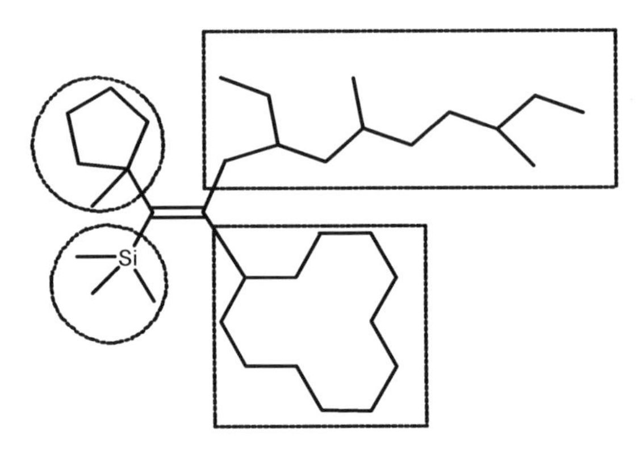

III. Now you must prioritize the substituents on each carbon as "high" or "low" using the priority rules.

1. first consider the atom directly bound to the double bond carbon. Rank them by atomic number - higher atomic number, higher priority:

I > Br > Cl > S > F > O > N > C > H

If the atomic number is the same but the mass is different (comparing isotopes of the same atom), the heavier isotope is higher priority. Thus: $^{13}C > {}^{12}C > {}^{2}H$ (denoted D, for deuterium) $> {}^{1}H$

2. a. If the directly bound atoms are the same (A and B in the figure), move out to the highest priority atom (use atomic number) bound to A and B (A1 and B1 in the figure). **NOTE:** you ONLY consider the next atom out if the first atoms (A and B) are the same; otherwise they do not matter at all!!!!!

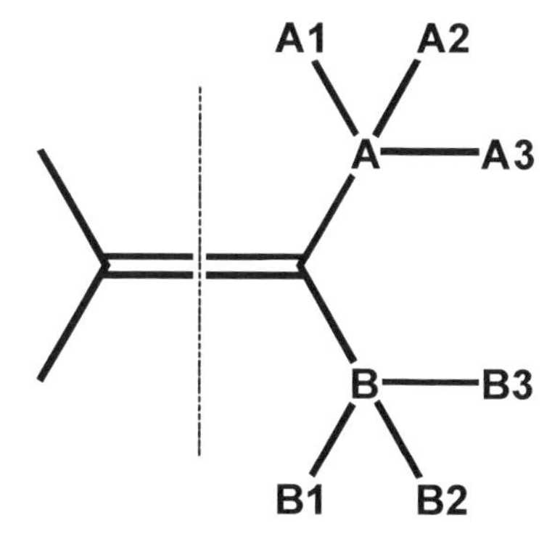

b. If A1 has higher priority than B1, then the "A" substituent has higher priority. If (and only if) A1 and B1 are ALSO tied (same atom), you have to move to the next-highest priority atom (A2 and B2) on A and B. If A2 is higher priority than B2, the "A" branch is the higher priority substituent. If A2 and B2 are tied, compare the next-highest priority substituents on A and B (A3 and B3). If A3 is higher than B3, the "A" branch is higher priority.

c. **If** A1, A2, and A3 match B1, B2, and B3, **then** you must begin comparing atoms attached to A1 and B1 using the same methodology described for comparing atoms on A and B, and continue moving farther from the initial attachment point in this manner until you find a difference between the A and B branches that allows you to make an assignment.

Examples:

$-C(CH_3)_3$ > $-CH(CH_3)_2$ > $-CH_2CH_3$ > $-CH_3$ > $-H$

$-CH_2F$ > $-CH(CH_3)_2$

$-OCH_3$ > $-O-H$

$-CH_2Br$ > $-CHCl_2$ > $-C(CH_3)_3$ > $-CH(CH_2Br)(CH_3)$

3. Multiple bonds are "broken open" to form the same number of single bonds to the same atoms. This method is sometimes called the "break and duplicate" approach; break the pi bonds, duplicate the atoms to which they were bound on each side of where the pi bond was:

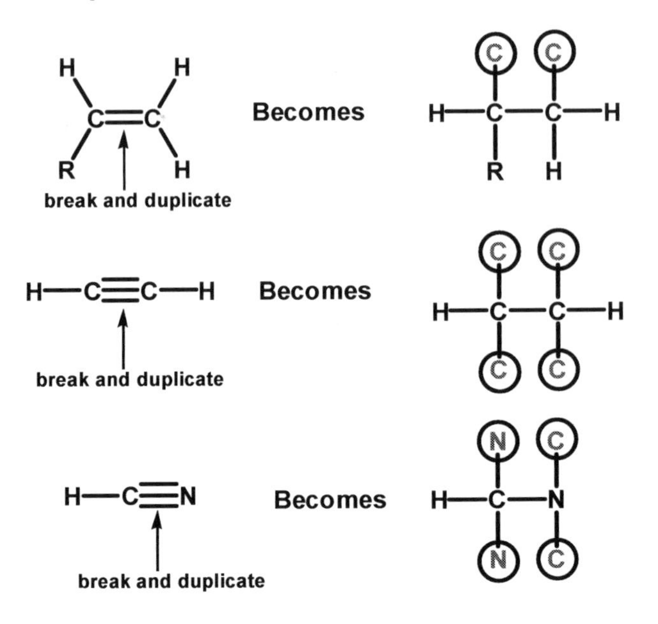

NOTE:
the circled atoms do not have any other atoms attached! They are imaginary constructs that serve only to help us prioritize, and have no other chemical significance!

Notes:

Priority Rules

Priority Rules:

First look at atom directly attached to the double bond;

(A) Higher Atomic # has highest priority

2. If same atomic number, higher mass = higher priority (D > H, ^{13}C > ^{12}C, etc.)
3. If the directly attached atoms are identical, move to the highest priority atom attached to the each of those atoms; the group substituted with higher atomic number substituent gets higher priority (basically apply rules 1 and 2 to next atom out, and so forth).

(B)

4. If a substituent is doubly or triply bonded to another atom, treat it as if it is two or three separate single bonds; **meaning:**

(C)

Notes:

Naming Epoxides

To name an epoxide, add the 'epoxy' prefix to the parent chain, and indicate the two carbon numbers to which the epoxide oxygen is bound:

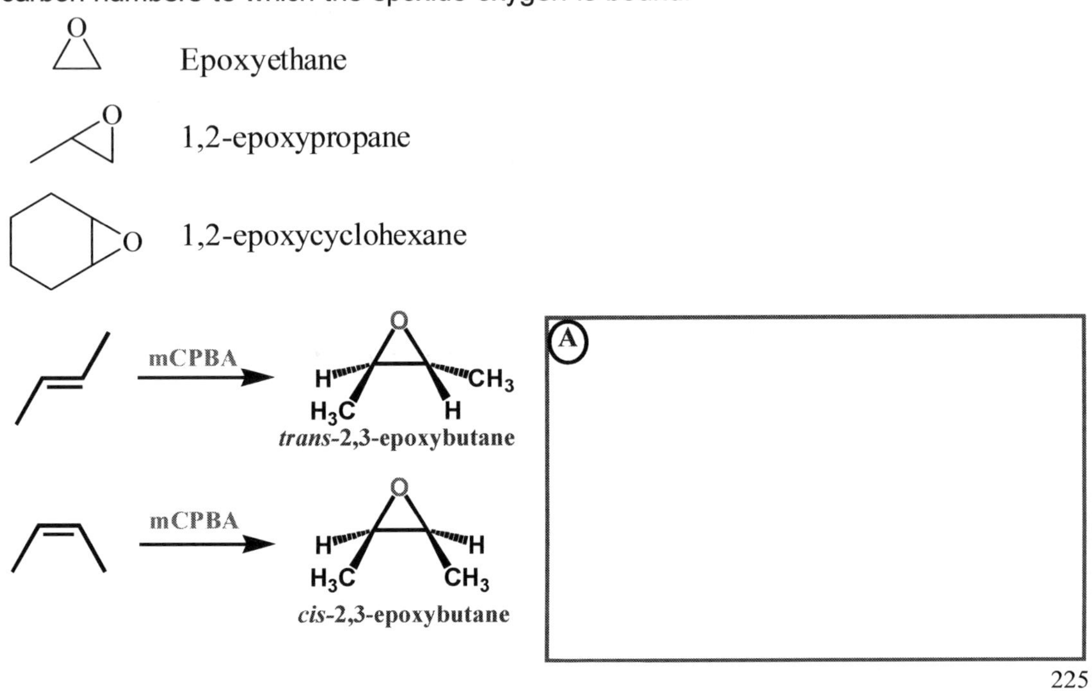

Notes:

Angle Strain

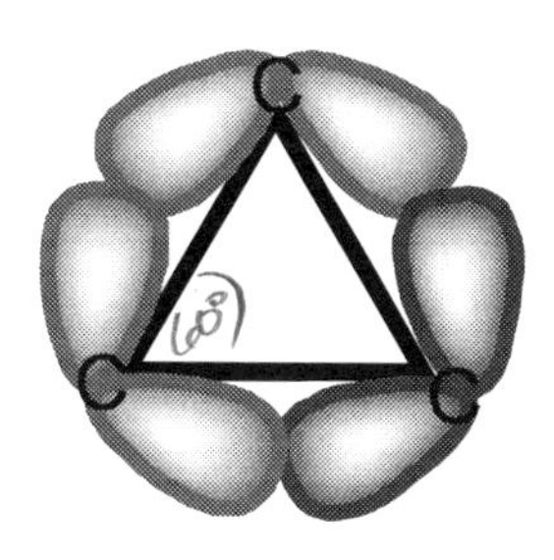

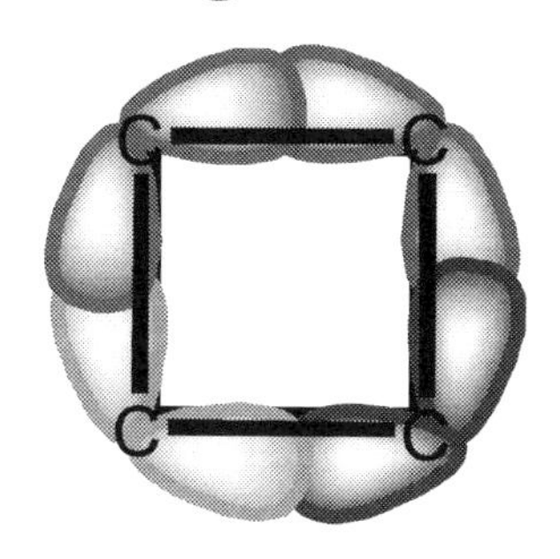

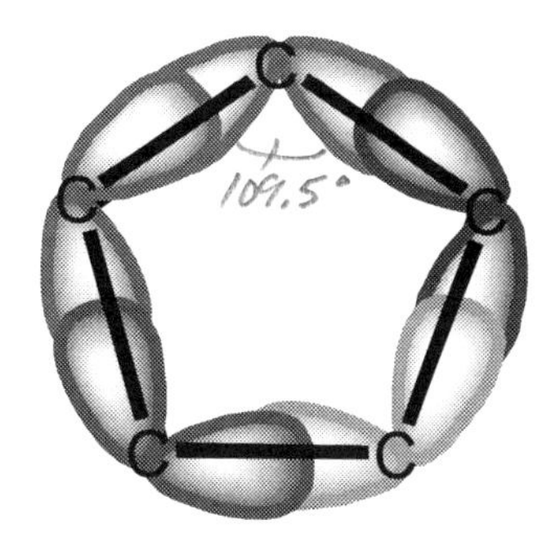

The **origin of angle strain** - poor orbital overlap

Good orbital overlap is needed for a strong bond

* Can't have 2 orbitals on the same atom close together or e^- repel each other *

Now we consider the other contributors to ring strain: steric interactions (and torsional strain)

18

Notes:

Chair conformer super stable! - It can twist to adjust angles to ≈ 109.5° (no angle strain).

- minimize steric strain b/c all substituents are staggered.

Observations:

Now that we can name alkenes, let us turn to some structure-property relationships.

1. more subbed = more stable

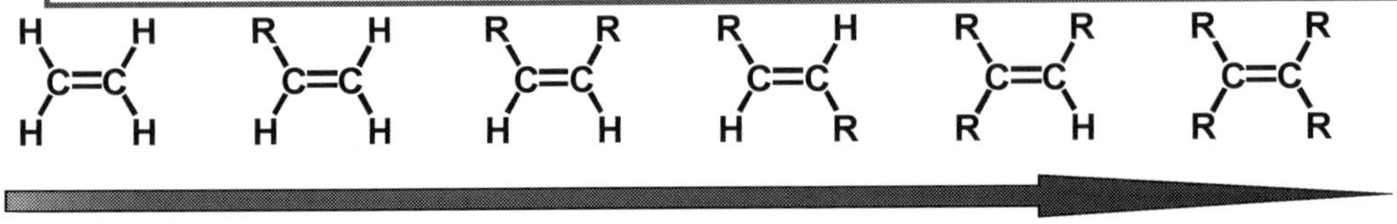

LESS STABLE → MORE STABLE

2. Trans-isomer > cis-isomer

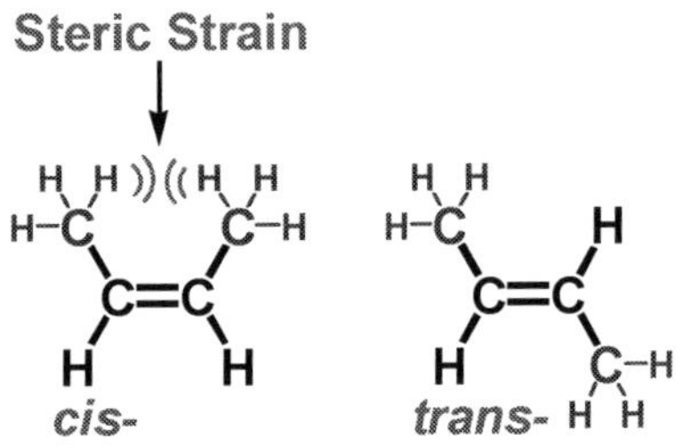

Steric Strain

cis- *trans-*

163

Notes:

"Zaitsev's Rule" - more stable alkene

"Hoffman Product" - less " "

Relationships

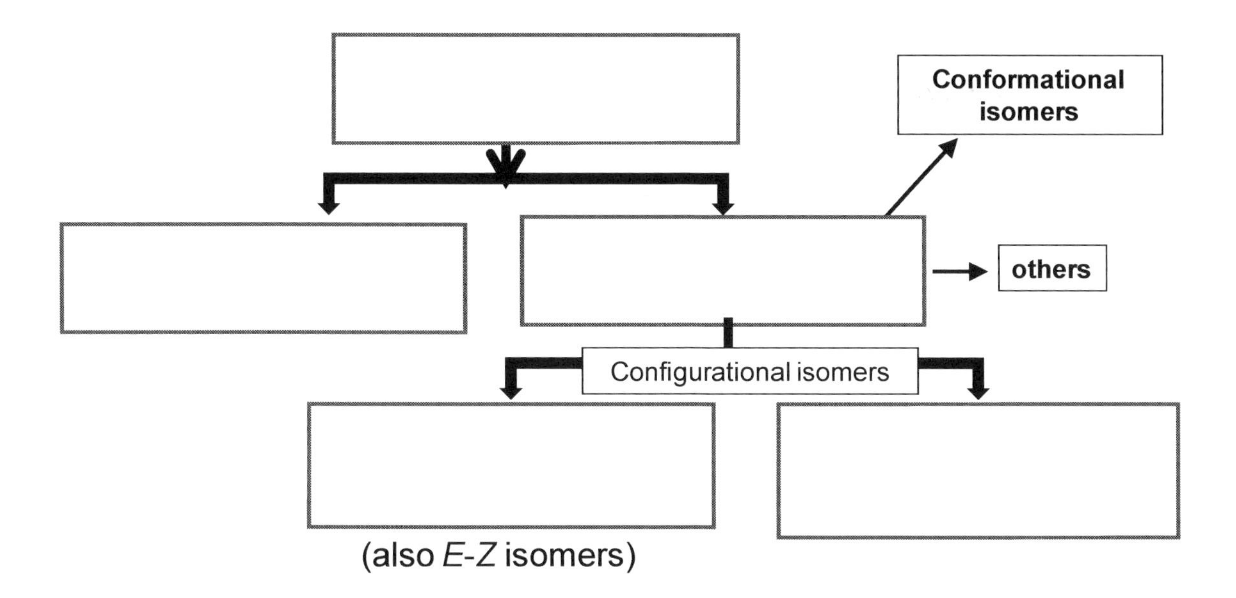

Notes:

Stereoisomers:

Ⓐ same atom connectivity, but different orientation of atoms in space

Conformational stereoisomers:

Ⓑ interconvertable by bond rotation

- •staggered and eclipsed conformers (Newman Projection)
- •*gauche* and *anti* conformers
- •Chair or boat chair conformers of cyclohexane

Configurational stereoisomers:

Ⓒ not interconvertible w/o bond breakage

**cis* and *trans* isomers **E* and *Z* isomers

NOW:

enantiomers:

Ⓓ chirality / non-superimposable mirror image

diastereomers:

Ⓔ non-superimposable non-mirror image

Notes:

Assigning Absolute Configuration: Newman

1. (A) Assign Priorities

   (see the "**Easy EZ**" section); priority rules are often referred to as the Cahn-Ingold-Prelog rules with reference to chirality.

2. (B) Point 4th priority away or to back

3. Determine the direction of procession of the other three groups, from **highest to lowest**:

   clockwise = (C) R

   counterclockwise = (D) S

CH_3CH_2–C(H)(OH)(CH_3) → Newman: CH_3 (3), H_5C_2 (2), OH (1) — (E) R-2-butanol

CH_3CH_2–C(H)(CH_3)(OH) → Newman: OH (1), H_5C_2 (2), CH_3 (3) — (F) S-2-butanol

Notes:

Fischer Projections

In addition to standard dash-wedge notation, there is another very useful manner in which to represent the 3D shape of a molecule called a **Fischer projection**:

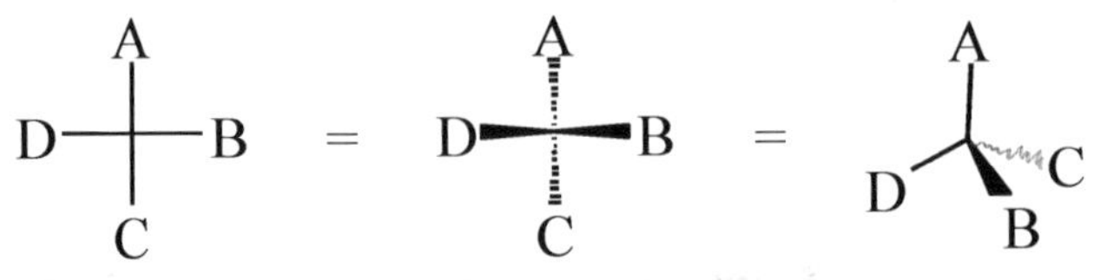

Ⓐ horizontal lines point toward
vertical lines point away

Ⓑ

46

Notes:

S_N1, S_N2, E1 and E2
Nucleophilicity versus Basicity

Nucleophilicity is a measure of how readily a compound (a nucleophile) is able to attack an electron-deficient atom

Nucleophilicity is measured by a rate constant (*k*).

(A) $k \propto E_a$ → rate is proportional to activation energy

UNLIKE

Basicity, which is a measure of how favorable it is for a compound (a base) to shares its lone pair, often with a proton, in solution.

Basicity is measured by the acid dissociation constant (K_a).

(B) $\Delta G = -RT \cdot \ln K$ → $\Delta G \propto \ln K$

Notes:

The Solvent Influences the Nucleophilicity

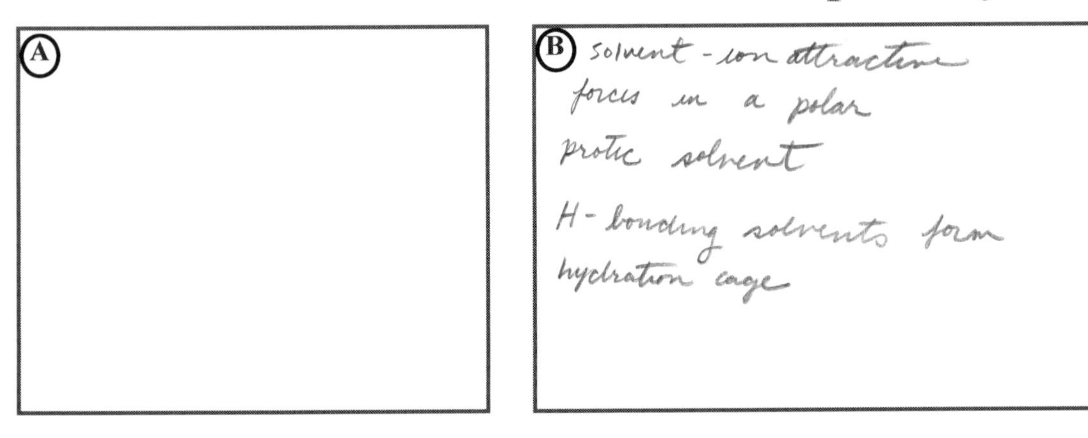

A
B
solvent - ion attraction forces in a polar protic solvent
H-bonding solvents form hydration cage

It is easier to break the ion-dipole interactions between a weak base and the solvent than between a strong base and the solvent

367

Notes:

- Don't use POLAR PROTIC if a strong $Nu^{\ominus}$
- Do use POLAR PROTIC if stable ion

The Solvent Influences the Nucleophilicity

Aprotic polar solvents such as DMSO facilitate the reaction of ionic compounds because they solvate ions very well. Examples of different classes of solvents:

nonpolar: hexane

moderately polar: diethylether, acetone

polar protic: H_2O, ROH

polar aprotic: DMSO (dimethylsulfoxide) $CH_3-S(=O)-CH_3$

Notes:

Nucleophile Trends

Nucleophilicity Trends

(A) Lots of charge = better $Nu^{\ominus}$

$O^{2-} > HO^- > H_2O$
$RO^- > ROH$
$HS^- > H_2S$

(B) EN = poorer $Nu^{\ominus}$

$H_2N^- > HO^- > F^-$
$H_3N > H_2O$
$RO^- > RCO_2^-$

(C) Size = better $Nu^{\ominus}$
Remember "BB" prefer E2 Rxn

$I^- > Br^- > Cl^- > F^-$
$HS^- > HO^-$
$PH_3 > NH_3$

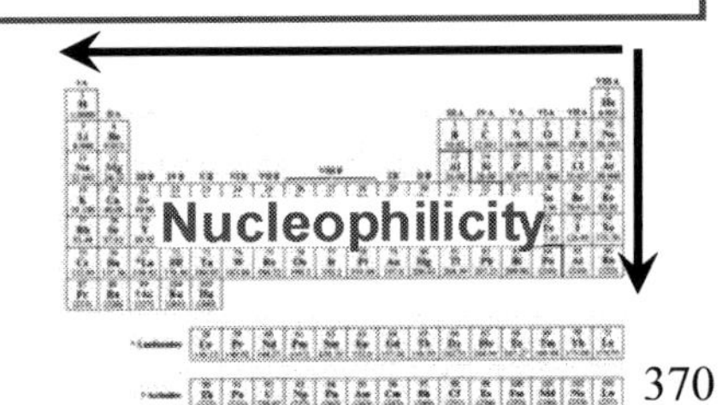

370

Notes:

Eliminate errors … accept no $S_N$ubstitute

S_N1 or E1 - No Rxn if 1°

depends on $Nu^{\ominus}$ or base — E2 or S_N2

strong $Nu^{\ominus}$/weak base — S_N2

strong Base/poor $Nu^{\ominus}$ - E2

strong $Nu^{\ominus}$/strong base — 1° S_N2
— 2° S&E
— 3° E2

The Rhett Smith Flowchart

Notes:

Substitution versus Elimination Problem

Example. Determine whether each reaction will proceed predominantly via S_N1, S_N2, E1, E2, or some combination thereof, and show the product(s)

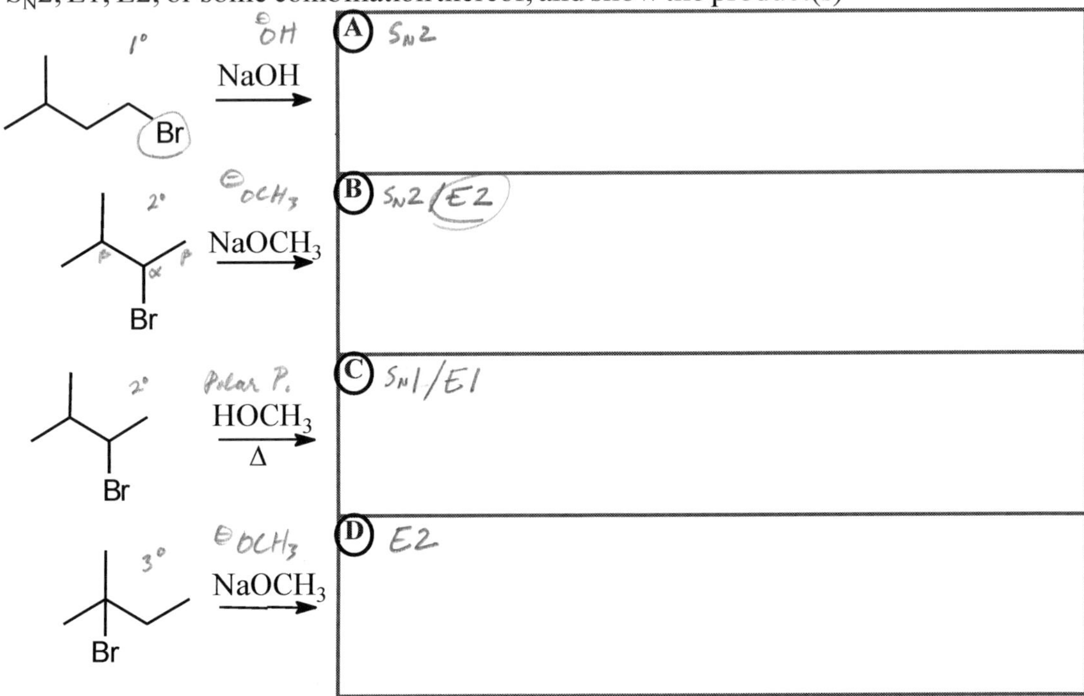

Notes:

Organic 1 Reaction Recap

In organic chemistry, we look at quite a few reaction mechanisms, some of which involve many individual steps to get from starting materials to the final product. It is worth noting that many reactions, even rather complicated ones, are composed predominantly of a few basic general mechanistic steps:

1. Coordination/Heterolysis (Lewis Acid / Base reaction):

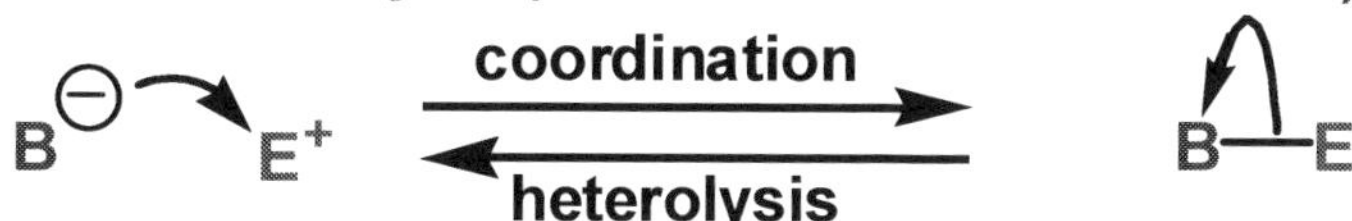

2. E2 Reaction:

LG
E2
B
H
+ BH + LG

3. Electrophilic Addition / Elimination:

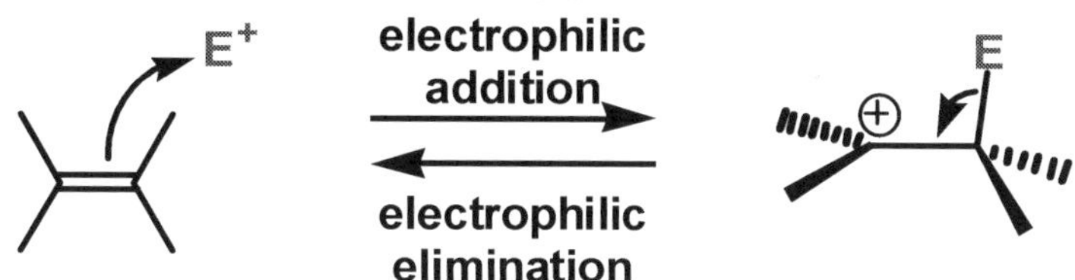

Notes:

Organic 1 Reactions Recap

4. Carbocation Rearrangement

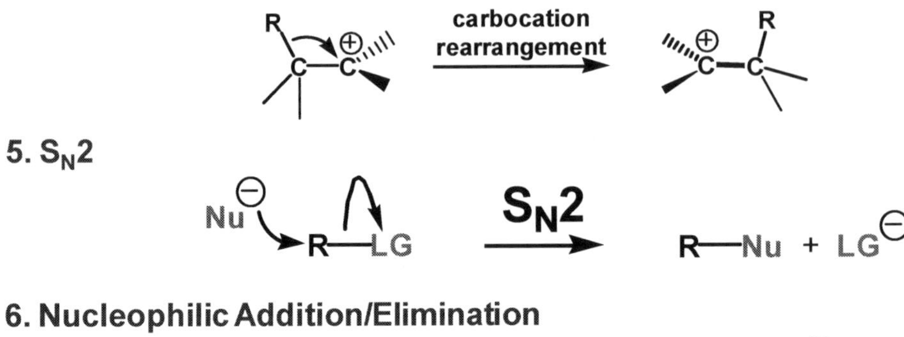

5. S_N2

6. Nucleophilic Addition/Elimination

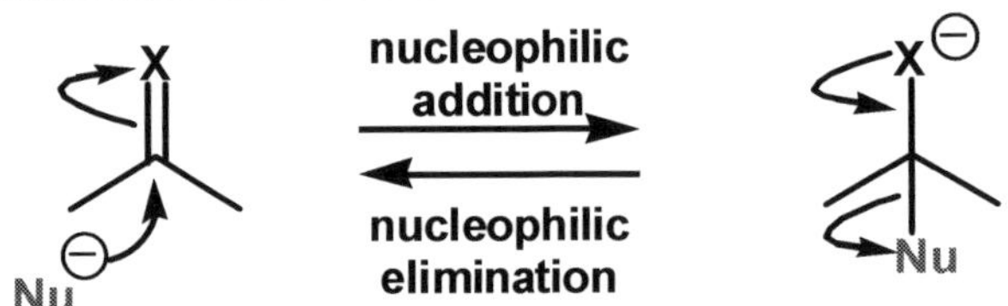

When designing or analyzing a reaction, evaluate the driving force for each step; electrons tend to move from electron rich sites to electron poor sites in order to reach more stable species.

Notes:

Lecture Set 1: CHAPTER 10 + 11
Reactions of Alcohols and Ethers and an Introduction to Organometallic Chemistry

Suggested Reading:

Suggested Problems:

Part I: Reactions of Alcohols

What reactions will we know from Organic 1 and Organic 2 that use an alcohol as a starting material?

A. Alcohol to Alkyl Halide
   1. By reaction with HX (via S_N1 or S_N2)
   2. Using $SOCl_2$, PX_3 (X = Cl, Br)

B. Alcohol to Alkene (dehydration via E1)

C. Alcohol reaction with sulfonyl chlorides to make sulfonate esters

D. Alcohol reaction with acid chlorides to make esters

E. Oxidation of alcohols to make aldehydes, ketones, or carboxylic acids

Notes:

A. OH $\xrightarrow{H-X}$ X

B. OH $\xrightarrow{H^+}$ (E1)

C.

Substitution Reaction: S_N1

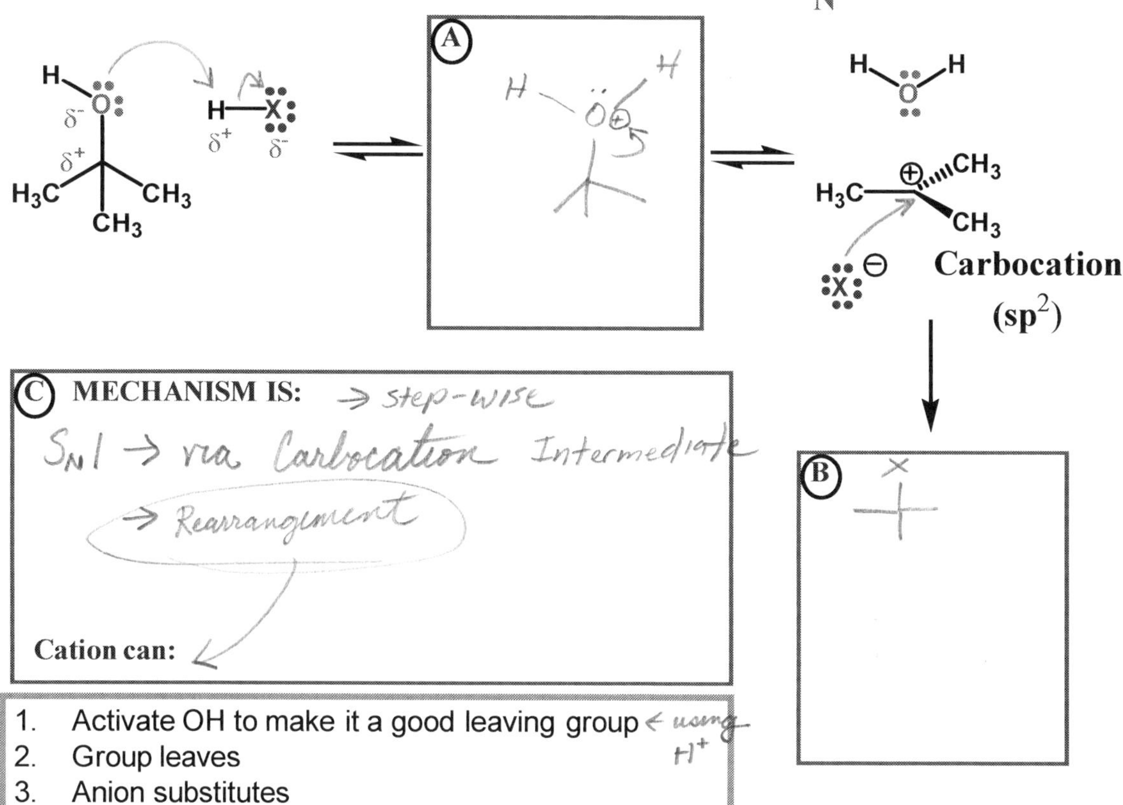

A
B
C
MECHANISM IS:
→ step-wise
S_N1 → via Carbocation Intermediate
→ Rearrangement
Cation can:
Carbocation
(sp^2)

1. Activate OH to make it a good leaving group ← using H^+
2. Group leaves
3. Anion substitutes

Notes:

- All strong bases are bad L.G. b/c they are unstable anions

The S_N2 Reaction Mechanism

Because (A) 1° Carbocation can't be formed

reaction between *primary* alcohol and HX proceeds by a mechanism

(B) concerted

Primary alcohols undergo the

(C) S_N2 Rxns

S-isomer $\xrightarrow{S_N2}$ R-isomer

(D) Inversion of stereochemistry → Walden Inversion

Transition State

1. Activate leaving group
2. S_N2 reaction

(recall: dashed line is a *partial* bond, i.e., less than two electrons, so C does NOT have 10 electrons here!

Notes:

Other Ways to make OH a good LG

Ⓐ PBr₃ makes R-Br from R-OH

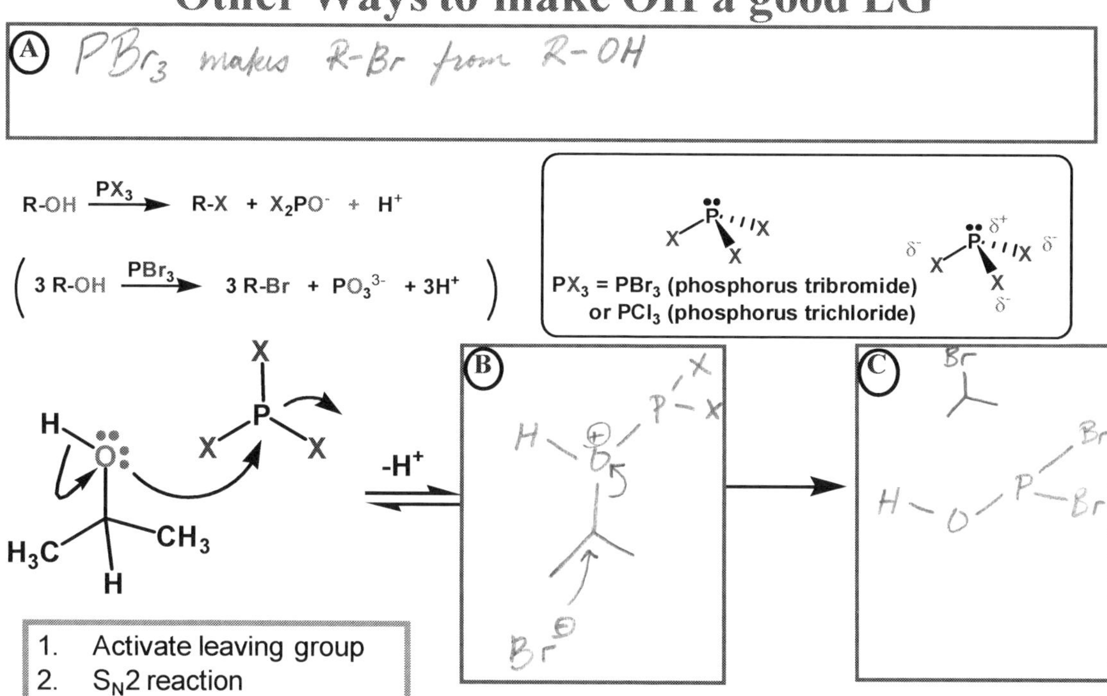

Adding base drives reaction by neutralizing the acid produced.
Note: 1 mole PX_3 can produce 3 mol RX

Notes:

Other Ways to make OH a good LG

HX is not the only reagent capable of transforming an alcohol into an alkyl halide. On the basis of the mechanism, any agent with a polar element-halogen bond that also activates OH to a good leaving group could work.

Ⓐ SOCl₂ makes R-Cl from R-OH

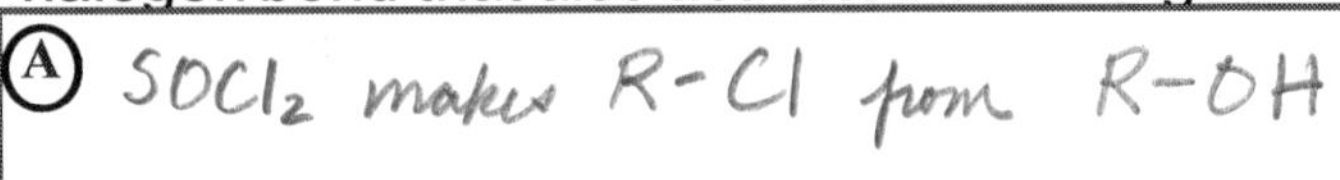

$SOCl_2$ = thionyl chloride

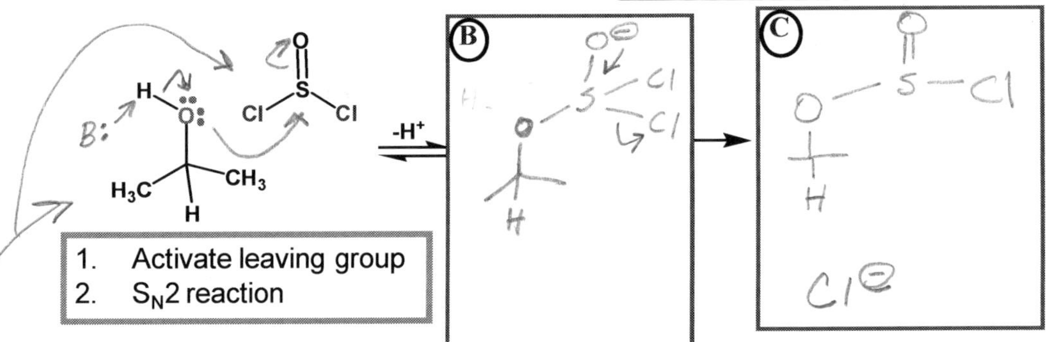

1. Activate leaving group
2. S_N2 reaction

Adding base drives reaction by neutralizing the acid produced.

Notes:

• π bonds break B4 σ bonds

Other Ways to make OH a good LG

a sulfonyl chloride

Now consider a very similar reaction where a sulfonyl chloride is reacted with an alcohol:

Pyridine

A

B

a sulfonate ester

both precipitate

Pyridine: A base

A sulfonate is an outstanding leaving group, so sulfonate esters are good starting materials for S_N2 reactions …

Notes:

- This Rxn is normally used w/ large excess of sulfonyl Chloride
- Rxn stops after sulfonate ester is formed

Some Common Sulfonate Esters

R—S(=O)(=O)—O⁻ ⟷ R–S(=O)(=O⁻)... ⟷ R–S(=O)(=O)–O⁻

In effect, converting the hydroxide group to a sulfonate ester makes it an ideal starting material for nucleophilic substitution:

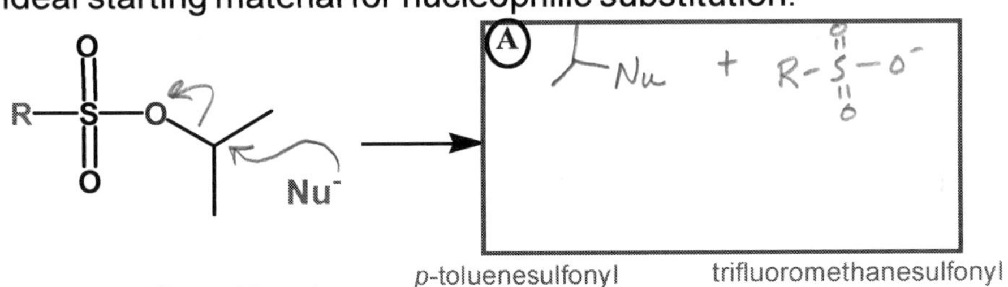

Two specific sulfonyl chlorides are often used to give particularly notable leaving groups:

p-toluenesulfonyl chloride

trifluoromethanesulfonyl chloride

F_3C—S(=O)(=O)—Cl

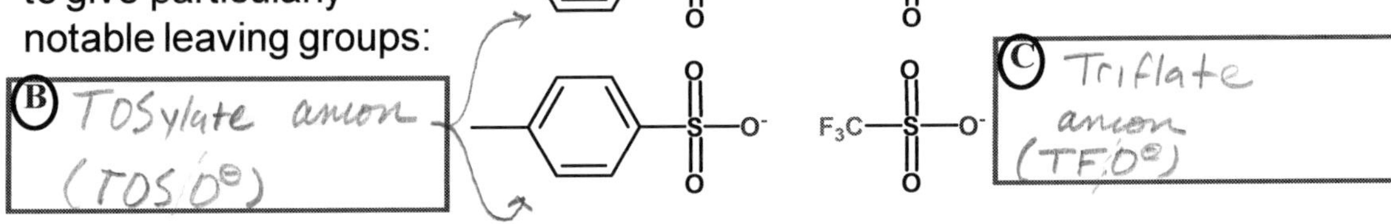

Notes:

Reaction of Alcohols with Acid Chlorides

Alcohols can react with acid chlorides just as they do with thionyl chlorides!

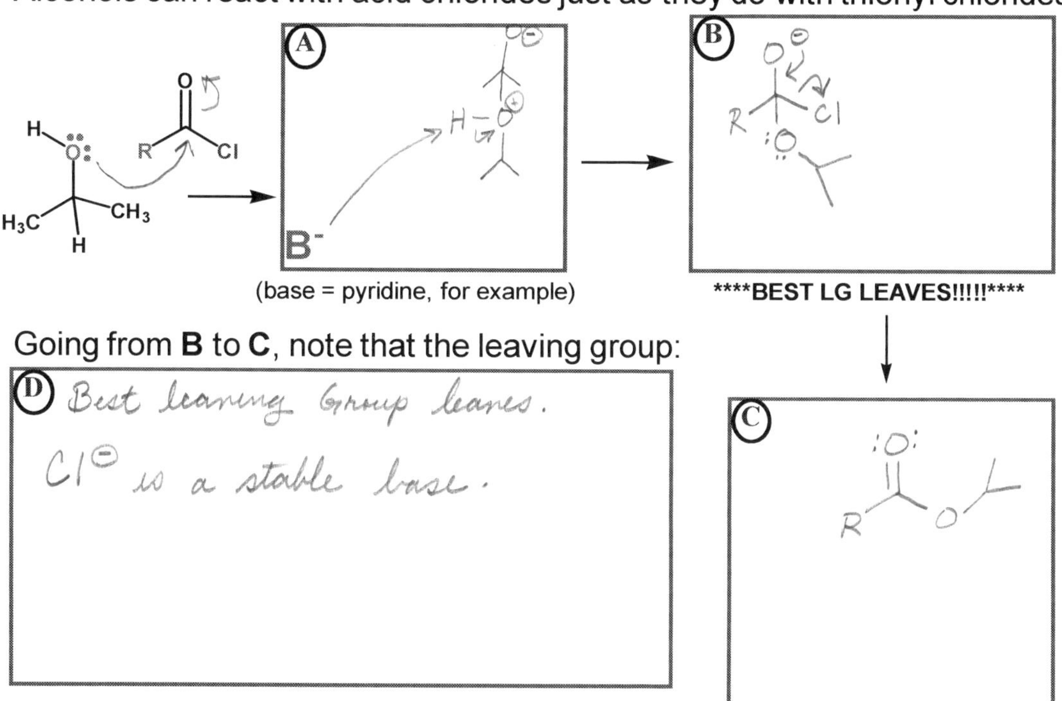

A
B
H
O
R
Cl
H_3C
CH_3
H
B⁻
(base = pyridine, for example)
****BEST LG LEAVES!!!!!****
Going from **B** to **C**, note that the leaving group:
D
Best leaving Group leaves.
Cl⊖ is a stable base.
C

an ester

Notes:

Problem: Esters and Sulfonate Esters

Example. Provide the reactant, reagent(s), or products to complete the reactions given below:

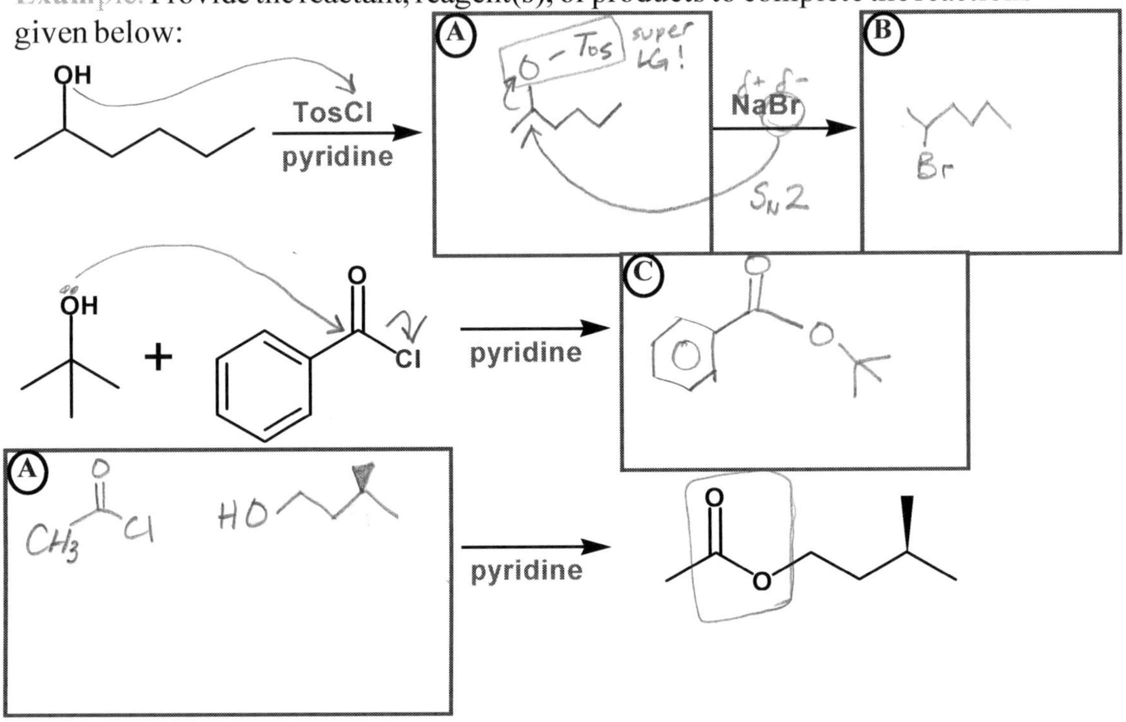

OH

TosCl

pyridine

A

O - Tos super LG!

NaBr

S_N2

B

Br

OH + Cl pyridine

C

A

CH_3 Cl HO

pyridine

O

O

Notes:

Oxidation and Reduction

$$CH_4 + 2\,O_2 \longrightarrow \text{(A)}\ CO_2 + H_2O$$

When methane burns in the presence to form carbon dioxide, it is said to undergo "oxidation". The opposite of oxidation is "reduction"

In general, the more C-H bonds that are replaced by C-O bonds in an alkane, the more oxidized the carbon atom becomes:

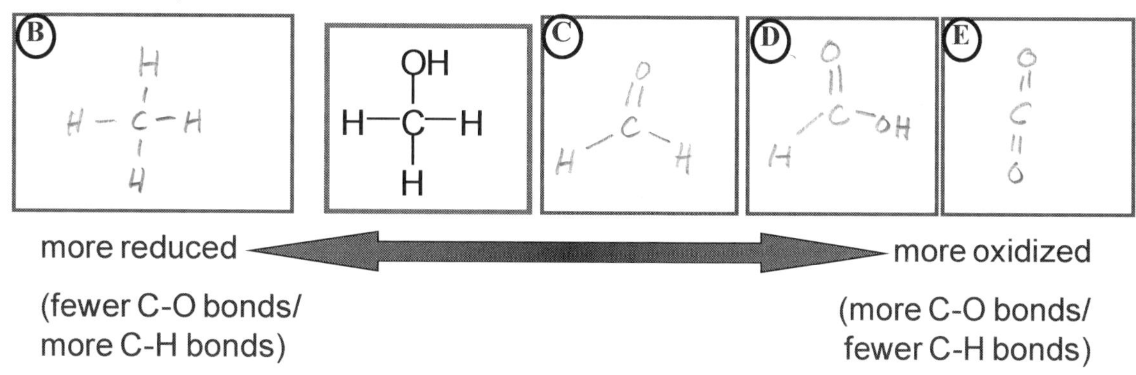

Notes:

Oxidation Number (≠ Formal Charge)

In order to quantify the concept of reduction/oxidation ("redox" for short) processes in organic chemistry, we can assign oxidation numbers to the atoms. The oxidation number of an atom depends upon the other atoms bound to it as follows:

1. Start at zero. For ***each bond*** to a more electronegative atom bound to an atom, add 1. For each bond to a more electropositive atom, subtract 1. For each identical atom, add 0.
2. Sum the individual numbers at each atom to get the oxidation number.

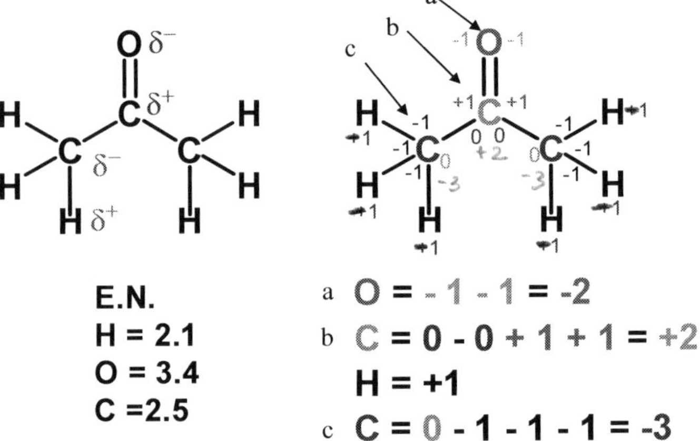

Notes:

Oxidation Number Problem

Example. Assign oxidation numbers to the carbons in these molecules.

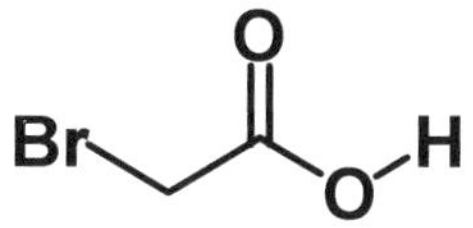

Br–CH₂–C(=O)–O–H

H–C≡N

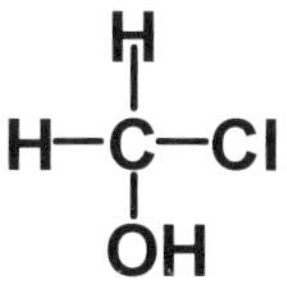

H–C(H)(Cl)–OH

Answer

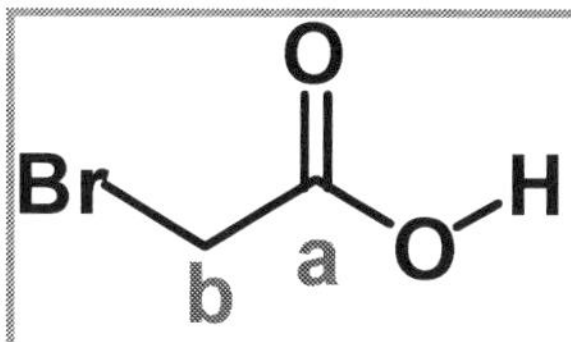

Br–CH₂(b)–C(a)(=O)–O–H	a: +1+1+1+0 = +3 b: +1-1-1+0 = -1
H–C≡N	-1+1+1+1 = +2
H–C(H)(Cl)–OH	-1-1+1+1 = 0

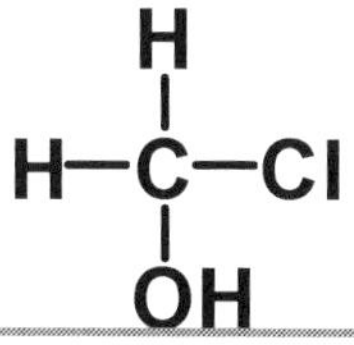

Notes:

Oxidation Number Problem 2

Example: Assign an oxidation number to each carbon in the following reactions (reactants and products). For each reaction, decide whether a net oxidation or reduction takes place.

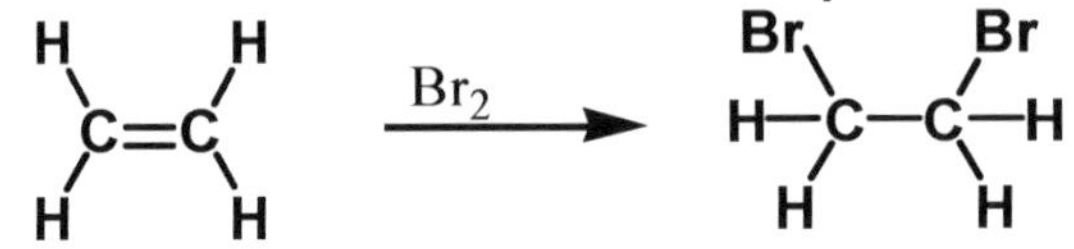

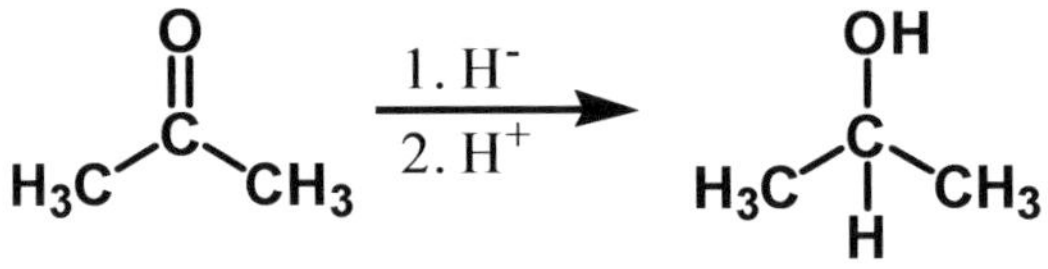

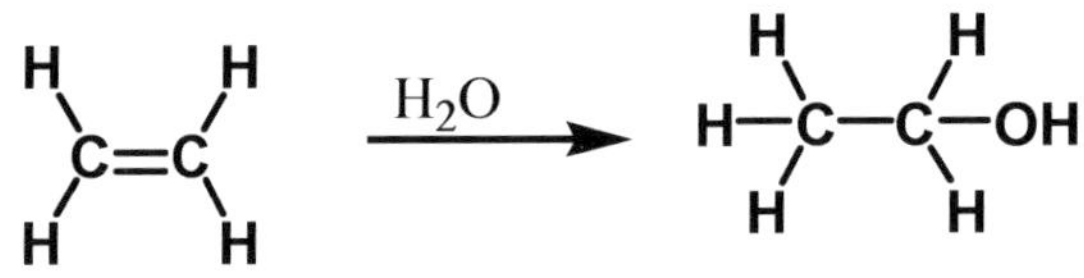

$$H_2C{=}CH_2 \xrightarrow{Br_2} H_2C(Br){-}C(Br)H_2$$

$$H_3C{-}C({=}O){-}CH_3 \xrightarrow[2.\ H^+]{1.\ H^-} H_3C{-}C(OH)(H){-}CH_3$$

$$H_2C{=}CH_2 \xrightarrow{H_2O} H_3C{-}CH_2{-}OH$$

Notes:

Oxidation Number Answer 2

The numbers in red refer only to carbon oxidation numbers.

$H_2C{=}CH_2$ (-2, -2) $\xrightarrow{Br_2}$ $BrCH_2{-}CH_2Br$ (-1, -1)

Oxidation: Each carbon changes from -2 to -1.

$(CH_3)_2C{=}O$ (-3, +2, -3) $\xrightarrow{1.\ H^-;\ 2.\ H^+}$ $(CH_3)_2CH{-}OH$ (-3, 0, -3)

Reduction: Central carbon changes from +2 to 0.

$H_2C{=}CH_2$ (-2, -2) $\xrightarrow{H_2O}$ $H_3C{-}CH_2OH$ (-3, -1)

No OVERALL oxidation or reduction, but one carbon is reduced (-2 to -3), one is oxidized (-2 to -1).

Notes:

Oxidation of Alcohols

Oxidation of alcohols is a useful way to make carboxylic acids, aldehydes, and ketones. The **oxidizing agents** typically used are transition metals in which the metal is in a high oxidation state; there is thus a driving force for the metal complex to oxidize something (the metal is concomitantly reduced).

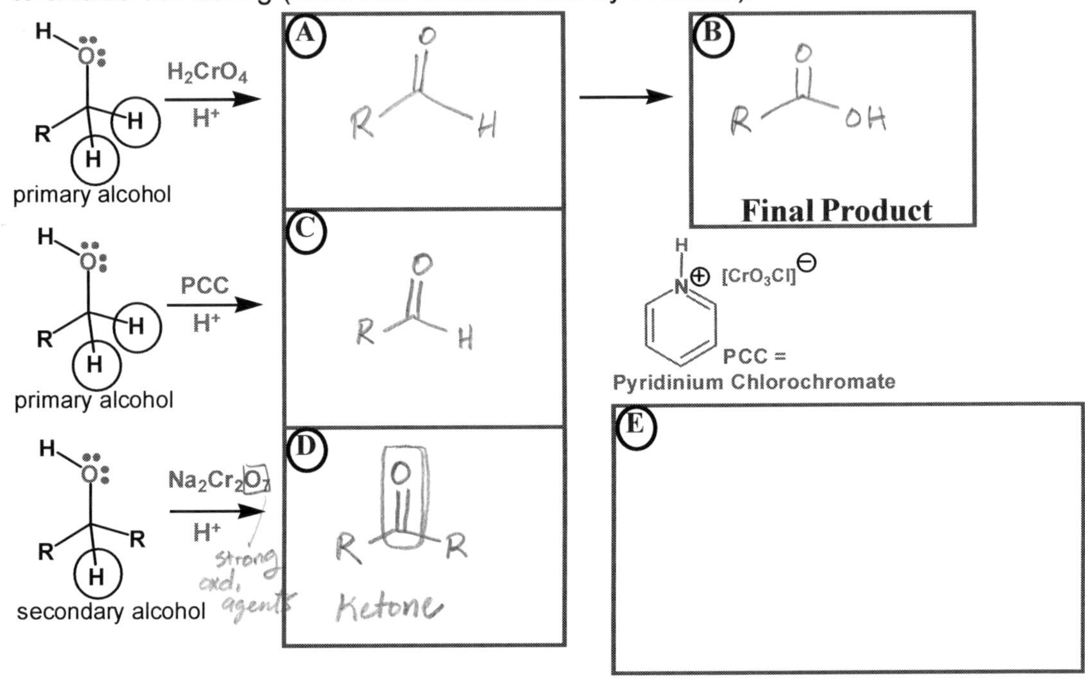

Notes:

Oxidation of Alcohols

The Mechanism of alcohol oxidation is fairly simple:

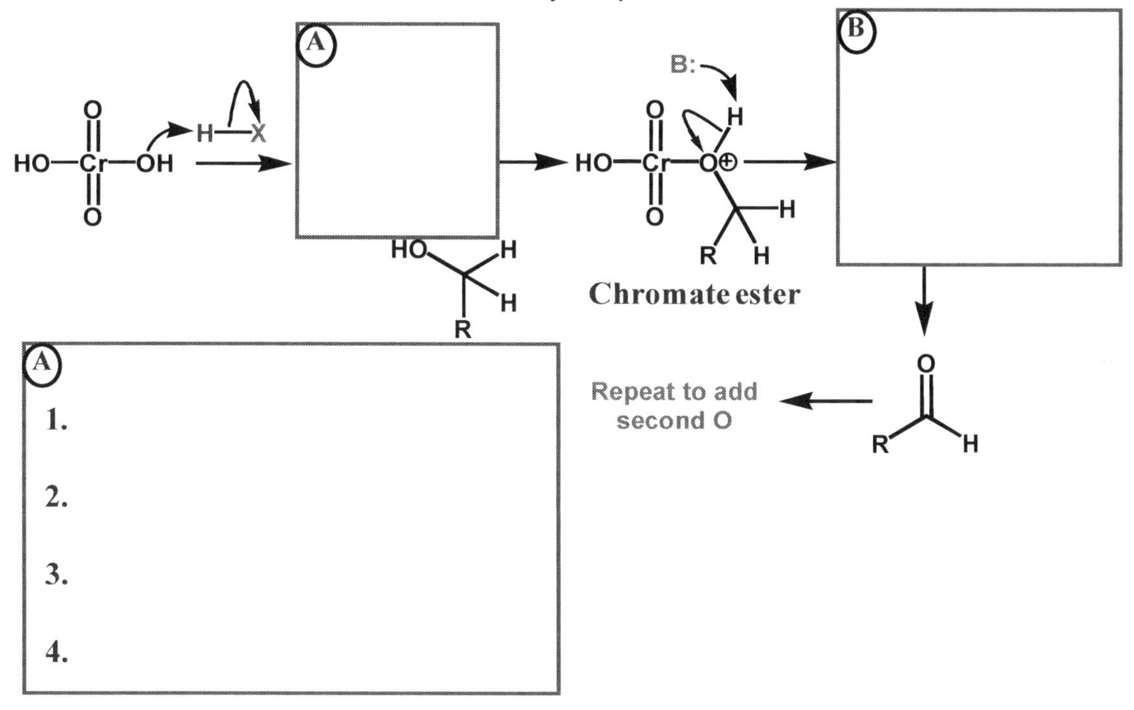

Notes:

Ring Opening of Epoxides

Epoxides are useful starting materials for nucleophilic substitution:

1.

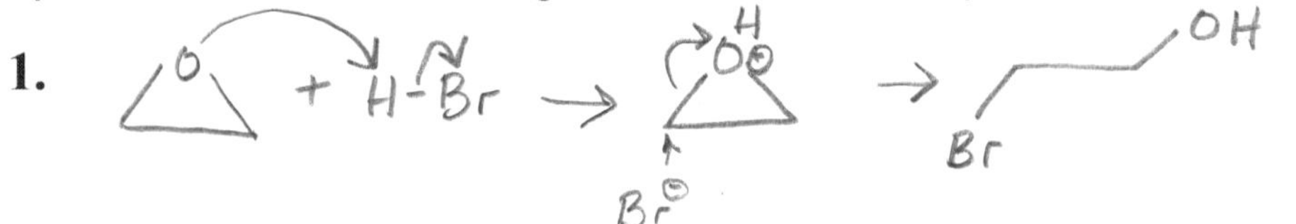

(A) Halohydrin

2.

H+

H_2O

OH

OH

OH

1,2-diol

Vicinal diols

(B)

Notes:

Mechanism Dictates Regioselectivity

Which carbon is attacked depends on the conditions / mechanism

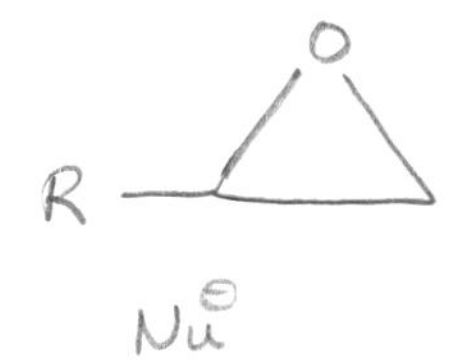

(A)

R

Nu⊖

Acid

H^+

S_N1-like Rxn

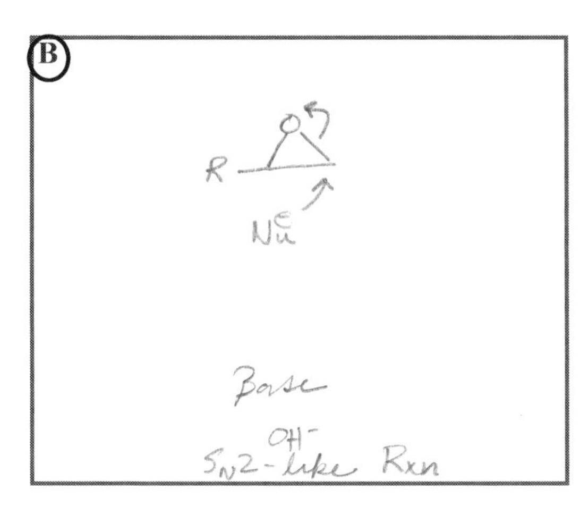

Notes:

Basic Conditions/Good Nucleophile

Under basic conditions in the presence of a good nucleophile, we have conditions suitable for an S_N2 reaction.

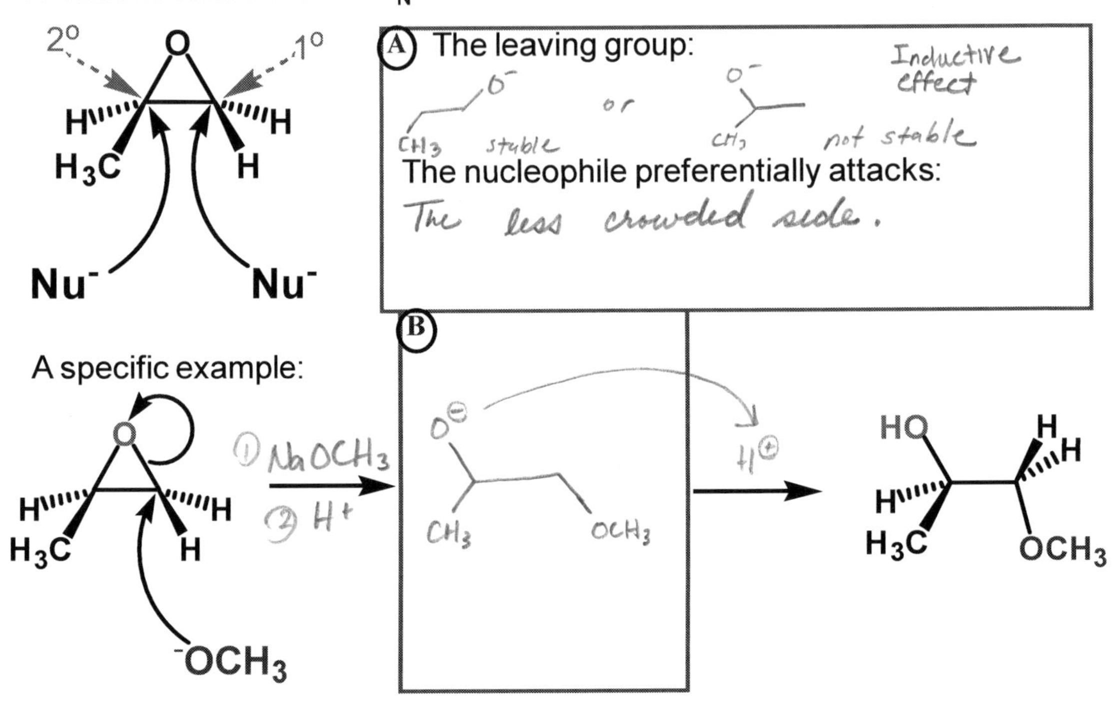

Notes:

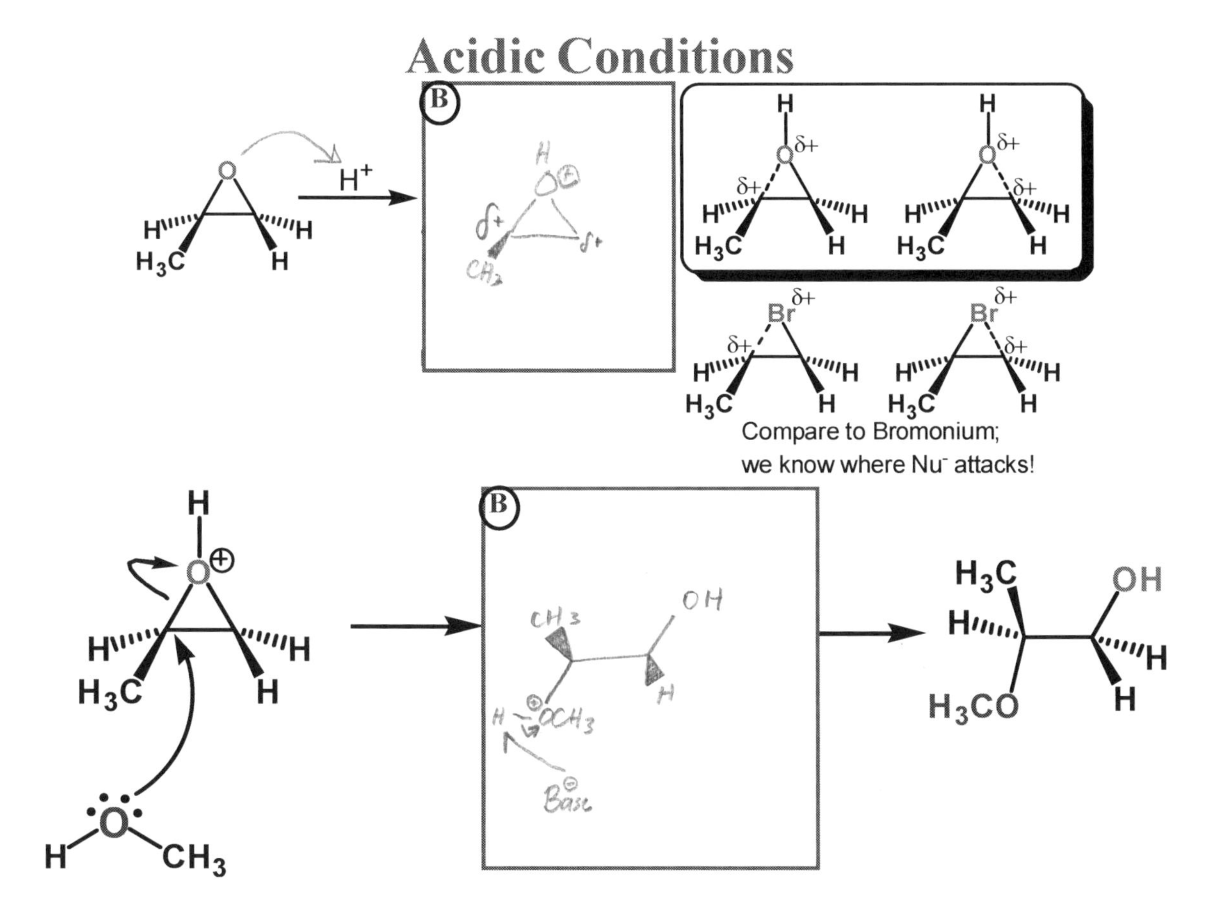
Acidic Conditions
H+
B
δ+
δ+
Br
Compare to Bromonium;
we know where Nu- attacks!
Base

Notes:

Stereochemistry

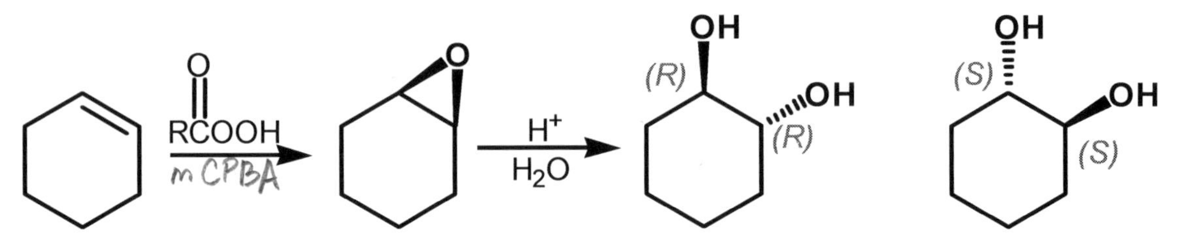

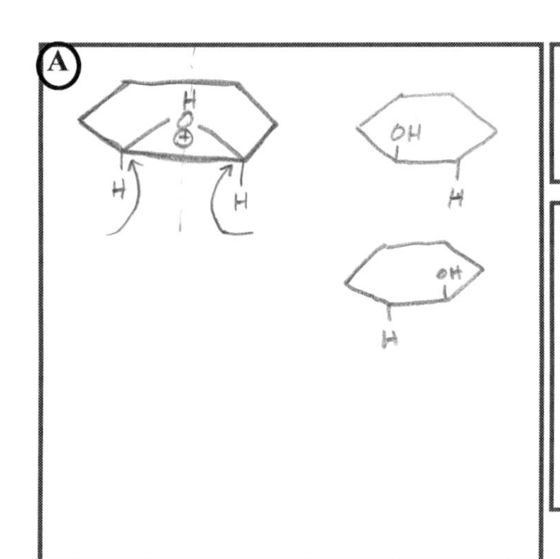

Addition is: Anti

Stereochemistry: Racemic mix

Notes:

Organometallic Compounds: C–M bonds

Epoxides can be used as starting materials to add additional carbon atoms if a carbanion source is used: (A) (B)

O; H; H; H_3C; H; $^{\ominus}CH_3$; H^+

$O^{\ominus}$ CH_3 | OH CH_3

A C–M (M = a metal) bond is very polar and places substantial negative charge on the carbon:

C–Li δ- δ+ C–Mg δ- δ+ C–Pd δ- δ+ δ- δ+ C–Cu

(C) C-M bond is organometallic
- C is Nu⊖
- &/or a strong Base

Notes:

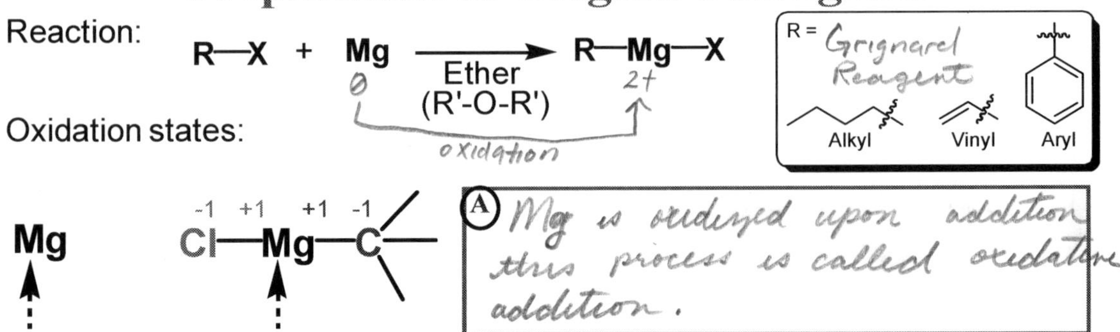

Preparation of Grignard Reagents

Reaction:

$$R{-}X \;+\; \underset{0}{Mg} \;\xrightarrow[\substack{(R'{-}O{-}R') \\ \text{oxidation}}]{Ether}\; R{-}\underset{2+}{Mg}{-}X$$

R = *Grignard Reagent*

Alkyl Vinyl Aryl

Oxidation states:

Mg

$$\overset{-1\;\;+1}{Cl}{-}\overset{+1\;-1}{Mg}{-}C$$

(A) *Mg is oxidized upon addition this process is called oxidative addition.*

This is an example of a very general mechanistic step wherein a compound adds to a metal center with concomitant formal oxidation of the metal center. This step is called:

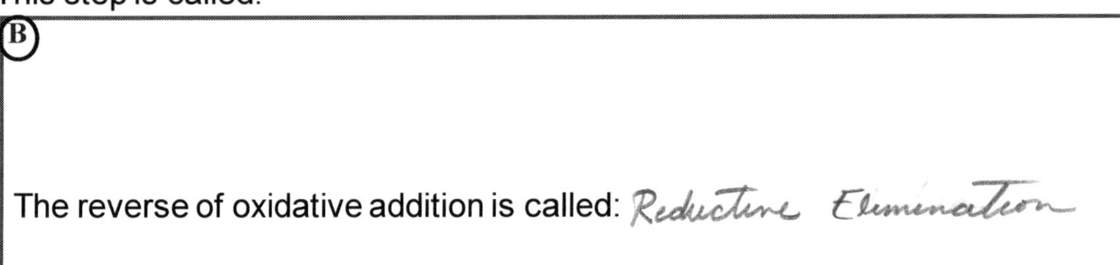

B
The reverse of oxidative addition is called: Reductive Elimination

Notes:

Reactions of Grignard Reagents

1. As a base: Grignard Reagent is much stronger than ^{-}OH.

$R-\ddot{O}H + Cl-Mg-CH_3 \rightarrow CH_4 + RO^{\ominus}\ {}^{\oplus}MgCl$

Grignard Reagents will deprotonate more acidic groups, i.e.:

Ⓐ $R-OH$

2. As a nucleophile: $R^{\ominus}-{}^{\oplus}MgX$ + $R^{\ominus}$ ⟶ LG → R–

strong Nu

weak Nu

Ⓑ

3. Exchanging groups with another metal:

$2R-Mg-Cl + CdCl_2 \longrightarrow CdR_2 + 2\,Cl-Mg-Cl$

extremely dangerous

Trading organic pieces between two metals is often called

Ⓒ Transmetallation

Notes:

Gilman Reagents (R_2CuLi)

Preparation from RLi:

2 Li + CuI →

A

Reaction with RX:

CuLi + Br →

B

Not a simple S_N2

Notes:

- "R" on Gilman Reagent replaces LG
- New C-C bond
- Gilman Reagents always replace LG w/ "R"

Gilman Reagents (R_2CuLi)

Works on alkenes and arenes!:

Me_2CuLi +

Br

A

Me

Me_2CuLi +

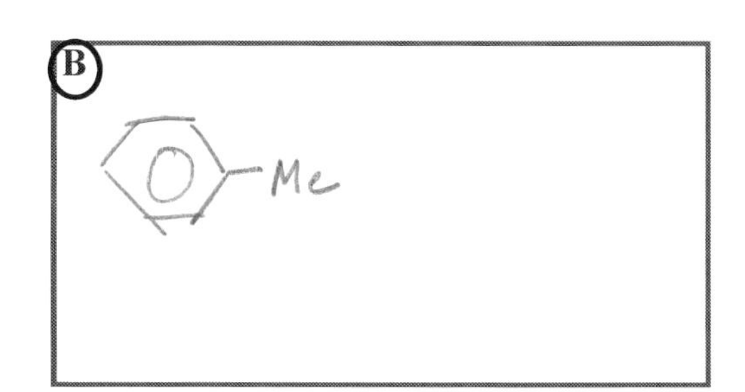

Notes:

- S_N2 needs LG to be on sp^3 <u>not</u> sp^2 hybridized "C"
- Gilman Reagent doesn't use S_N2 much.

Gilman Reagents (R_2CuLi)

...as well as allylic sites and adjacent to a carbonyl!:

Me_2CuLi + Cl

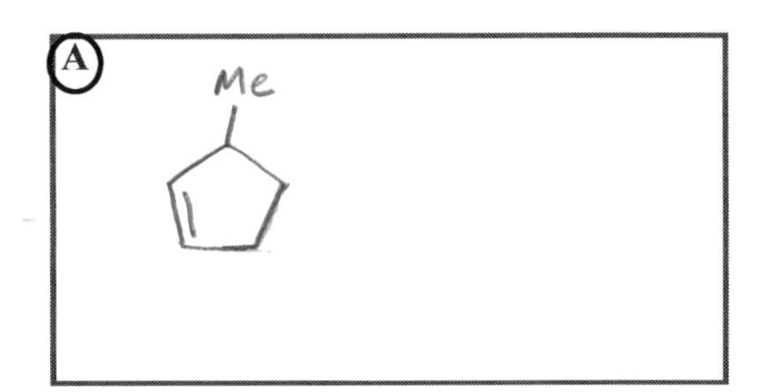

Me_2CuLi + O Br

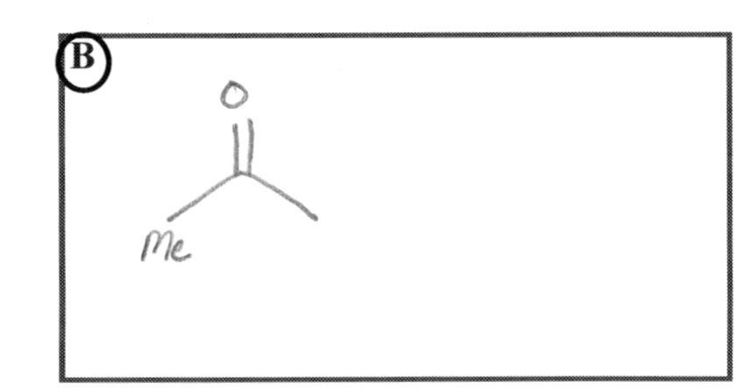

Notes:

- "Me" = CH_3

Pd-Catalyzed Coupling

Transition metal catalyzed reactions, particularly those employing Palladium (Pd) complexes, are growing ever more important and powerful for the formation of C-C bonds. A generalized Pd-catalyzed C-C bond forming reaction is given below:

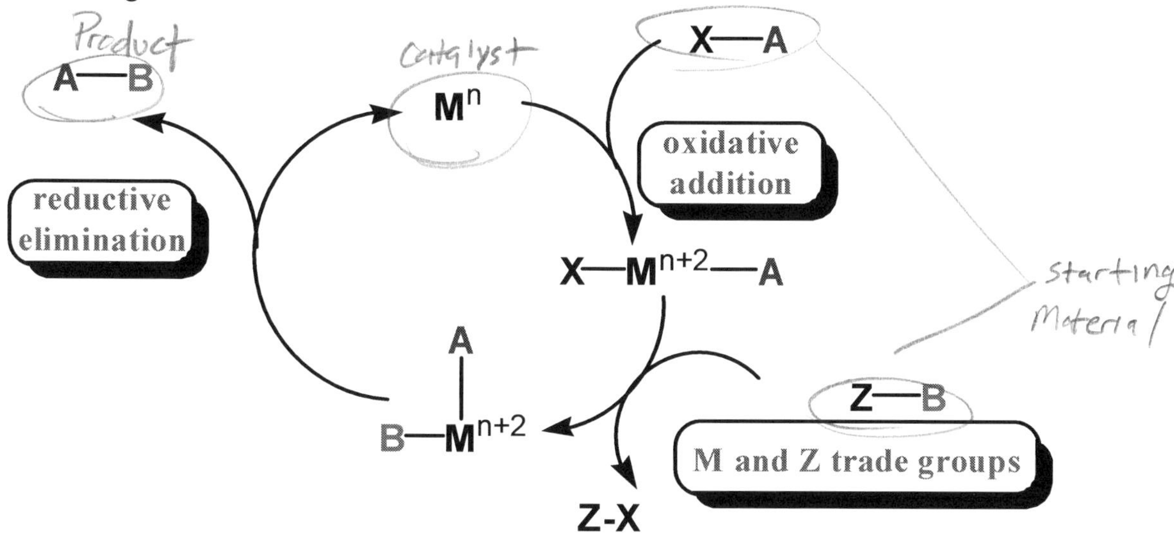

Notes:

Pd-Catalyzed Coupling: The Heck Reaction

The **Heck Reaction** couples:

(A) R–X + Alkene

R = aryl, benzyl or vinyl; X = OTf, Cl, Br, or I

$$R{-}X + H{-}CH{=}CH{-}R' \xrightarrow[N(C_2H_5)_3 \text{ (Base)}]{Pd(PPh_3)_4 \text{ (catalyst)}} R{-}CH{=}CH{-}R'$$

$H_3CO{-}C_6H_4{-}OTf$ (Triflate) + H–CH=CH–C_6H_5 $\xrightarrow[N(C_2H_5)_3]{Pd(PPh_3)_4}$

(B) coupling Rxn

$H_3CO{-}C_6H_4{-}CH{=}CH{-}C_6H_5$

* notice trans molecule

2-iodo-$C_6H_4{-}C(=O){-}OCH_3$ + $CH_2{=}CH_2$ $\xrightarrow[N(C_2H_5)_3]{Pd(PPh_3)_4}$

(C) 2-vinyl-$C_6H_4{-}C(=O){-}OCH_3$

Notes:

- PPh_3 = $\ddot{P}(C_6H_5)_3$

Pd-Catalyzed Coupling: The Heck Reaction

A simplified mechanism is:

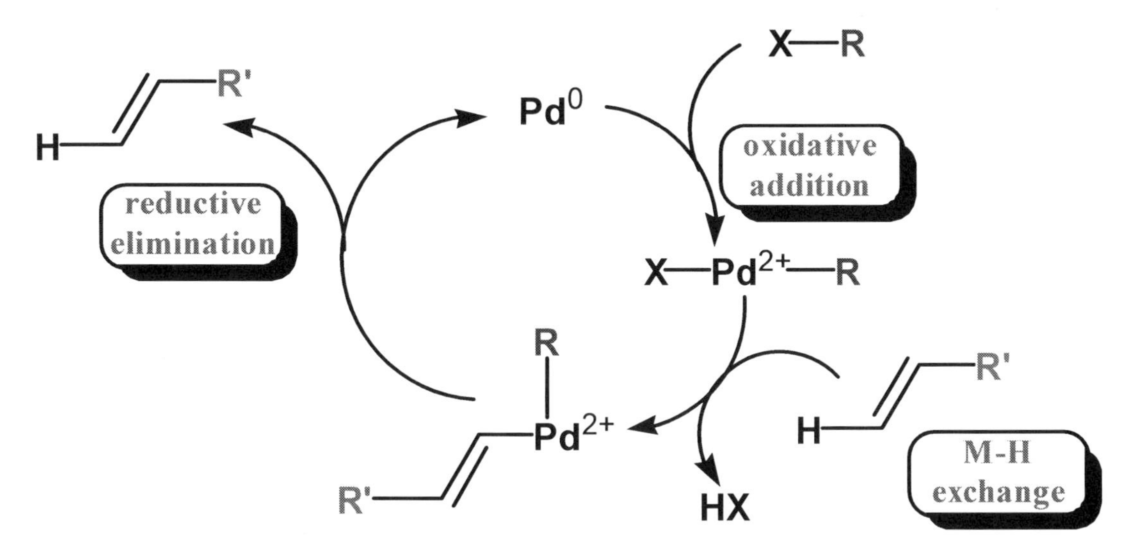

Notes:

Pd-Catalyzed Coupling: Stille Reaction

The **Stille Reaction** couples:

Ⓐ R-X + Sn Reagent   Sn = Tin = Stannous

R = aryl, benzyl or vinyl; X = OTf, Cl, Br, or I

$$R{-}X + R'{-}SnBu_3 \xrightarrow[\text{THF}]{Pd(PPh_3)_4} R{-}R'$$

uses non "Bu" LG

H_3CO–C_6H_4–OTf + $PhCH{=}CHSnBu_3$ $\xrightarrow[\text{THF}]{Pd(PPh_3)_4}$ Ⓑ H_3CO–C_6H_4–CH=CH–Ph

methyl 2-iodobenzoate ($C(=O)OCH_3$) + $PhSnBu_3$ $\xrightarrow[\text{THF}]{Pd(PPh_3)_4}$ Ⓒ methyl biphenyl-2-carboxylate (OCH_3)

Notes:

- Sn^{4+} is Toxic
- Stille can couple non alkenes unlike Heck Rxn

Pd-Catalyzed Coupling: Stille Reaction

A simplified mechanism is:

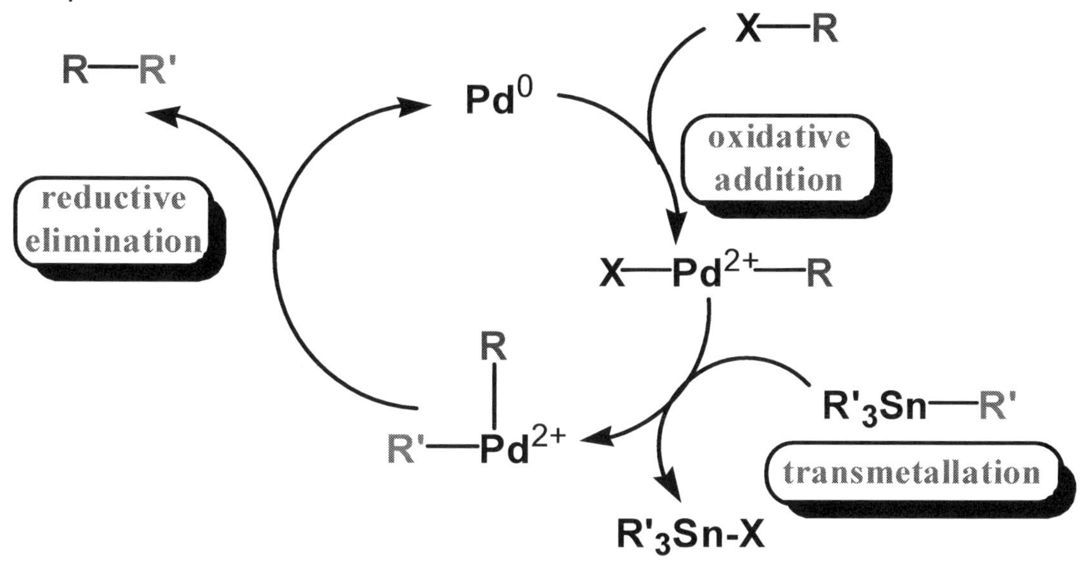

Notes:

Pd-Catalyzed Coupling: Suzuki Reaction

The **Suzuki Reaction** couples:

(A) R-X + Boron Reagent

R = aryl, benzyl or vinyl; X = Cl, Br, or I

$$R{-}X + R'{-}B(OR)_2 \xrightarrow[NaOH]{Pd(PPh_3)_4} R{-}R'$$

H_3CO–C₆H₄–CH=CH–Br + (catecholato)B–Ph $\xrightarrow[NaOH]{Pd(PPh_3)_4}$ (B) H_3CO–C₆H₄–CH=CH–C₆H₅

Base

methyl 2-iodobenzoate (OCH_3, I) + PhB(OH)$_2$ $\xrightarrow[NaOH]{Pd(PPh_3)_4}$ (C) methyl 2-phenylbenzoate (OCH_3)

25

Notes:

- may see interesting compounds bonded to the O on B

Pd-Catalyzed Coupling: Suzuki Reaction

A simplified mechanism is:

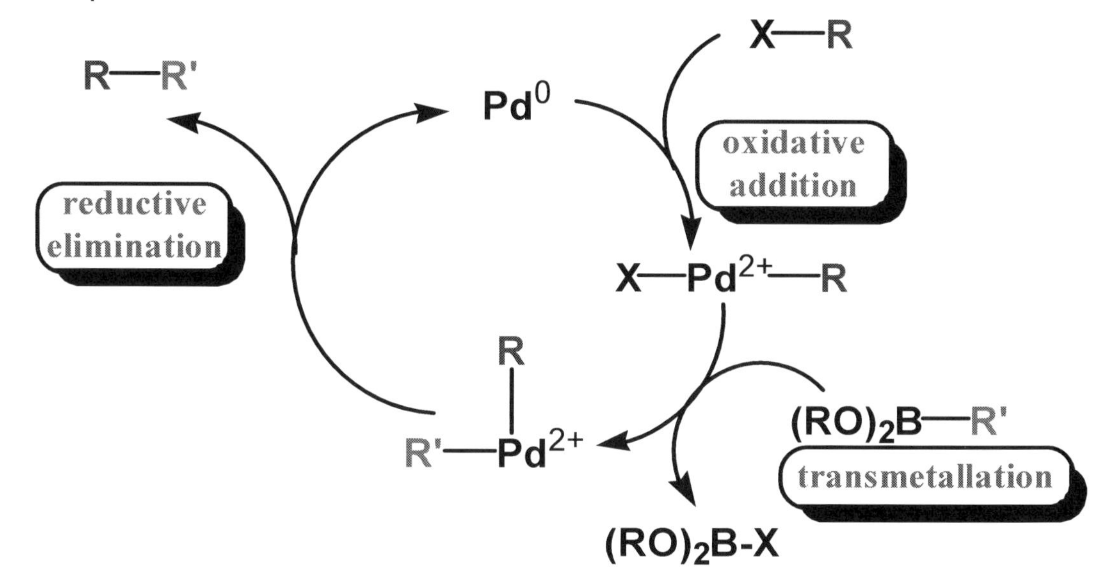

Notes:

Lecture Set 1 Checklist

- ❑ Be Able To:
  - ❑ Do things listed in previous checklists
  - ❑ Provide reagents, starting materials, or products needed to prepare sulfonate esters or esters from alcohols, and draw arrow-pushing mechanisms for these reactions.
  - ❑ Assign oxidation numbers to atoms in any given molecule and given a reaction determine which atoms are reduced and which are oxidized.
  - ❑ Provide reagents, starting materials, or products for the oxidation of alcohols to carboxylic acids, aldehydes, or ketones.
  - ❑ Provide reagents, products, or staring materials for ring-opening of epoxides under acidic or basic conditions, and draw arrow-pushing mechanisms for these reactions.
  - ❑ Provide a chemical equation showing how a Grignard Reagent is prepared.
  - ❑ Identify a given reaction as an example of oxidative addition, reductive elimination, transmetallation, or M-H exchange.
  - ❑ Provide reactants, reagents, or products for Pd-catalyzed coupling reactions discussed in this lecture set (Heck Reaction, Stille Reaction, and Suzuki Reaction).

By 2/15 (Exam 1) you should be a master of Lecture Set fundamentals.

Attend review sessions for more pointers!

Notes:

Lecture Set 2: Mass Spectrometry and Infrared Spectroscopy

skip pg. 93, 94
100 - 109

Suggested Reading:

Suggested Problems:

Identification of Organic Compounds

All previous chapters have focused primarily on understanding the structure and reactivity of organic molecules. Our abilities to determine the structure of a given compound and to reveal the products of a reaction are predicate upon techniques that reveal characteristics of the material such as:

Ⓐ Mass Spectrometry (molecular weight + its pieces)

Ⓑ IR Spectroscopy (Elemental composition)

Ⓒ NMR (Functional Groups)

Ⓓ (connectivity b/w pieces)

In this and the next lecture set, we will study some techniques that allow us to probe these features, as well as how said techniques work.

Notes:

* Paranthesis go in box A-D not techniques *

Technique 1: Mass Spectrometry

Analogy:

Ⓐ Fragments molecule into different pieces

What MS can reveal:

molecular mass:

Ⓑ How heavy each piece is
(Highest mass = whole molecule)

molecular formula:

Ⓒ What fragments go together

Structural features of the compound:

Ⓓ Way the molecule fragments (tells what elements are there & what functional groups)

Notes:

Mass Spectrometer

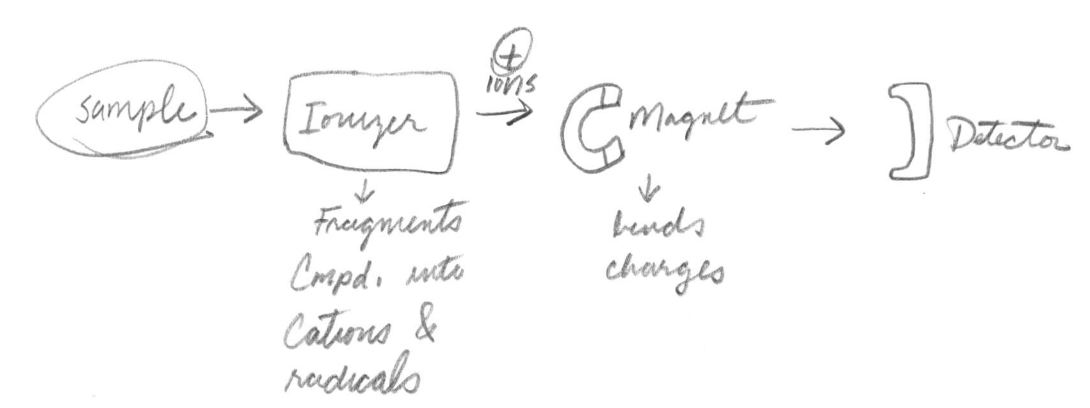

1) inject
2) cations
3) Magnet only deflects charged particles (heavier particles deflect less while lighter ones deflect more) (tells mass of ion)

Notes:

MS Shows Cations

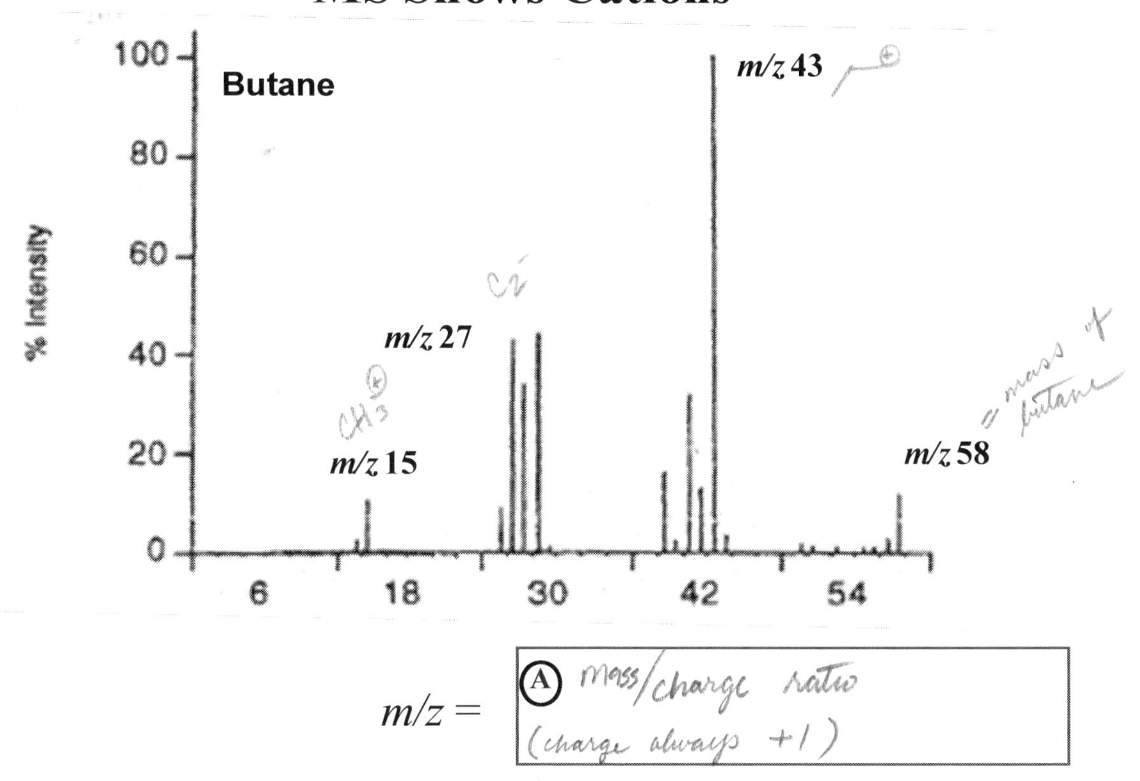

$m/z =$ Ⓐ mass/charge ratio (charge always +1)

Notes:

• **Nominal molecular mass:**

Ⓐ mass you get by using most abundant isotope

• **Peak with the highest *m/z* value:**

Ⓑ "molecular ion peak"

• **Other peaks:**

Ⓒ "Fragment ions"

Notes:

Base peak:

Ⓐ "highest intensity peak"

Predicting fragments:

Ⓑ think about most stable ⊕ & radicals

Ⓒ Breaks weak bonds

Notes:

The base peak at 43 indicates C2-C3 bond breakage is favored in fragmentation

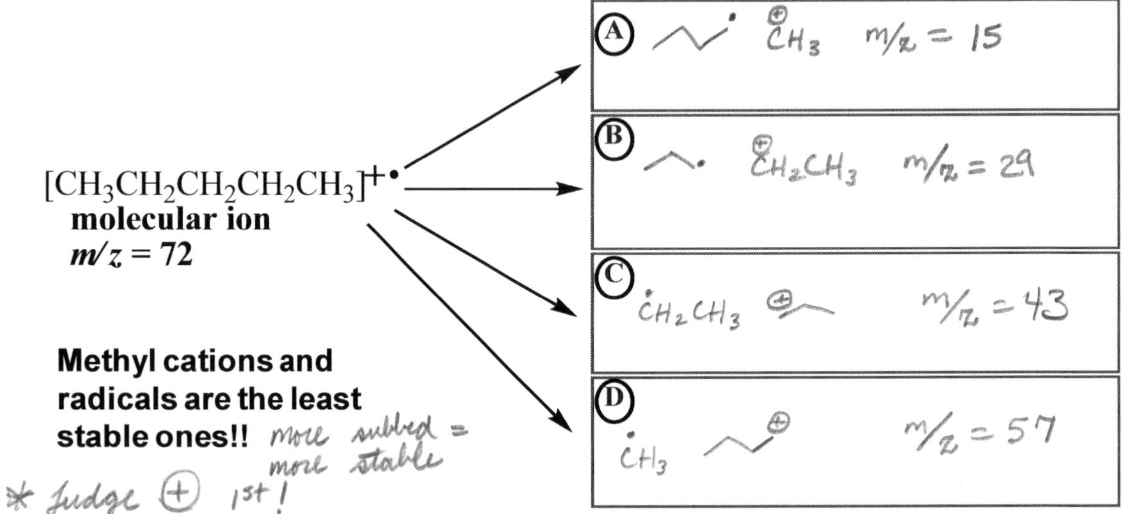

[CH3CH2CH2CH2CH3]+•
molecular ion
m/z = 72
A
CH3 m/z = 15
B
CH2CH3 m/z = 29
C
CH2CH3 m/z = 43
D
CH3 m/z = 57
Methyl cations and radicals are the least stable ones!!
more subbed = more stable
* Judge + 1st!

To identify fragment ions in a spectrum:

E

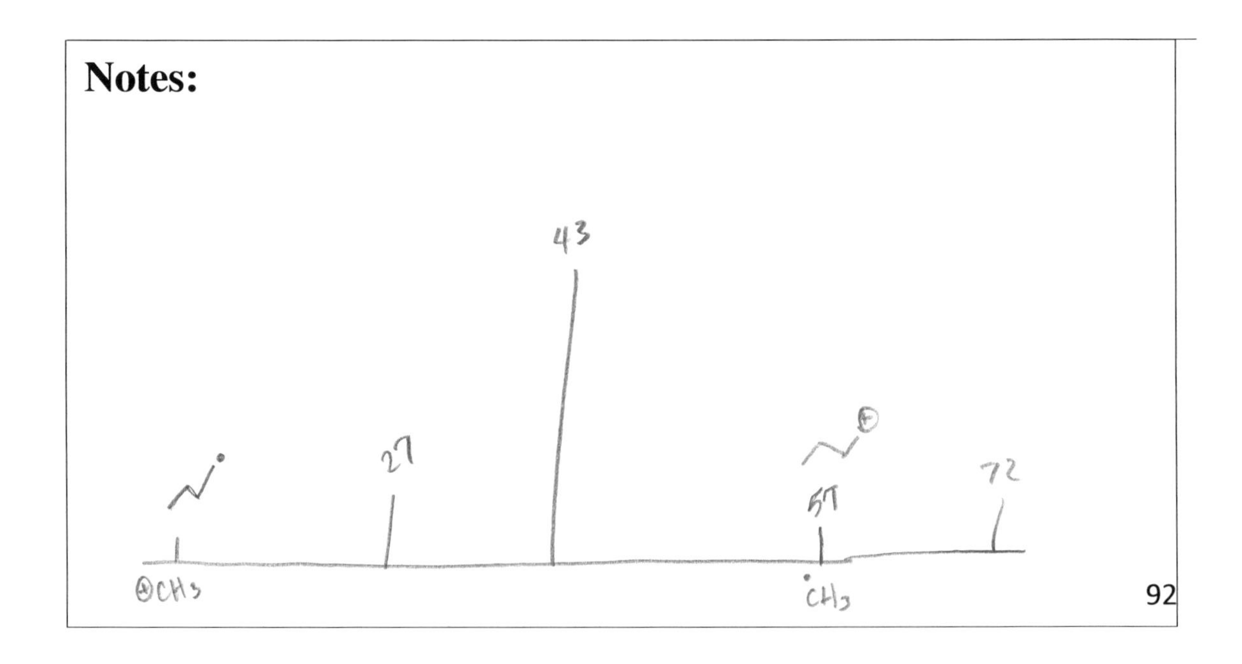

Notes:
43
27
57
72
CH3

Carbocations can Fragment

Once a cation is formed in the fragmentation process, it can still fragment to form new cations that do not form directly from the starting compound. Consider the molecular ion formed from propane:

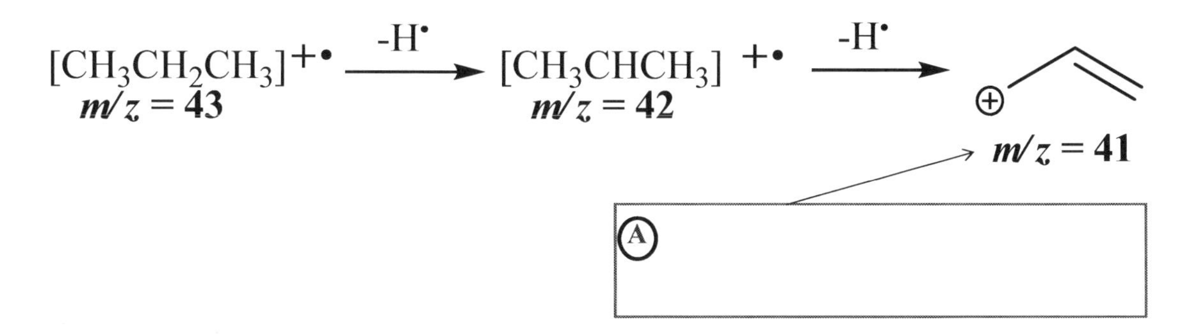

Notes:

Compounds with the same mass may be differentiated by their fragmentation pattern. For example, 2-methylpropane has the same *m/z* as butane, but intensities are different:

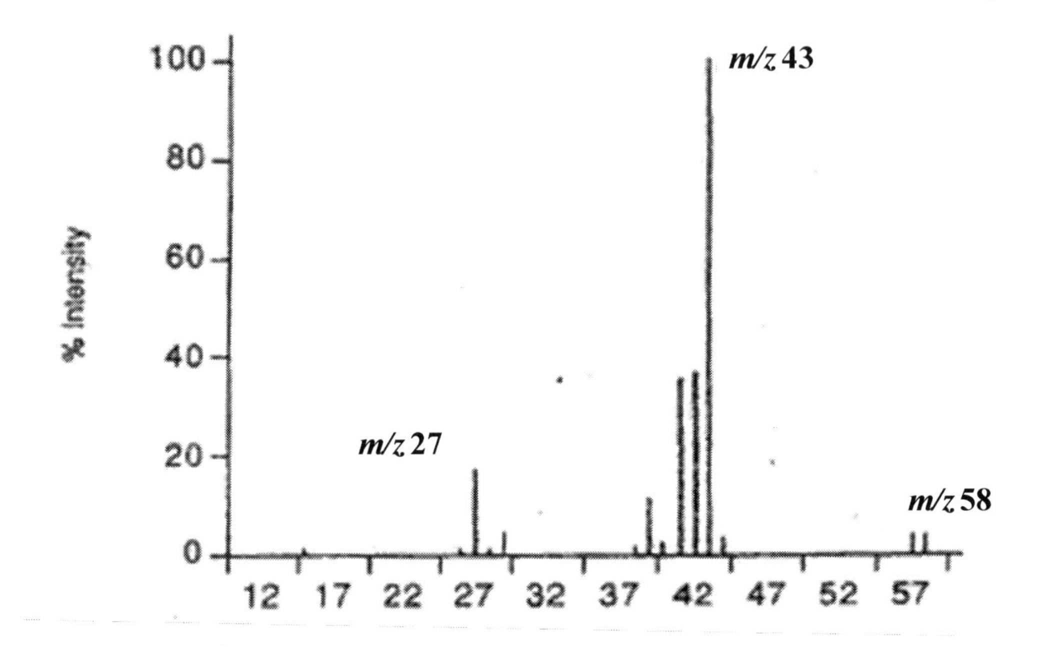

Notes:

Effect of Isotopes

Isotopic distribution patterns are quite useful:

For masses of ions

Ⓐ Single Isotopes → see different isotopes in mass spectrum

M + 2 peak

Ⓑ if M+2 = 1/3 height of "M" peak, then there is Cl in compound

if M+2 = same height of "M", then Br in compound

Large M + 2 peak

Notes:

- nominal mass (machine only reads most common isotope)

2-Bromobutane

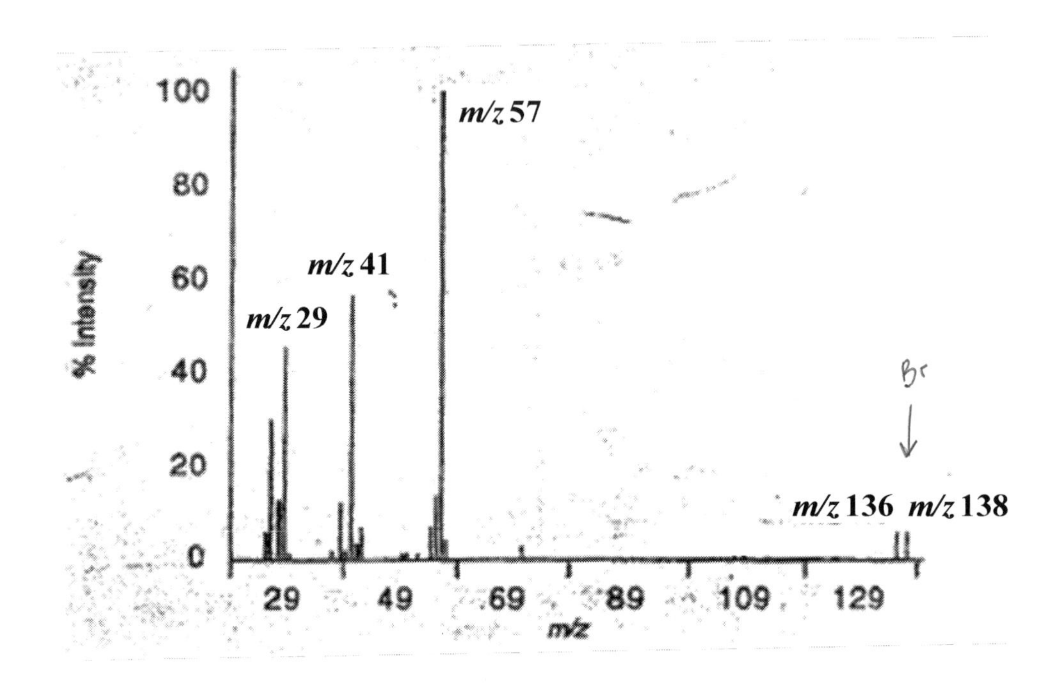

Notes:

- Br is usually the 1st fragment off

Predicting fragmentation:

(A) poor size = poor overlap = weak bond (1)
m/z = 138 & 136 = Br

Isotopic distribution for Br?

(B) m/z = 138 & 136 = Br

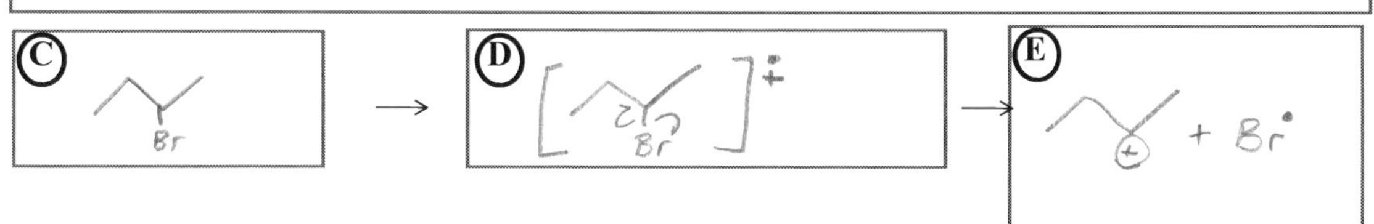

The initial cation:

(F) Butyl cation breaks up the same as in comp.

Notes:

2-Chloropropane

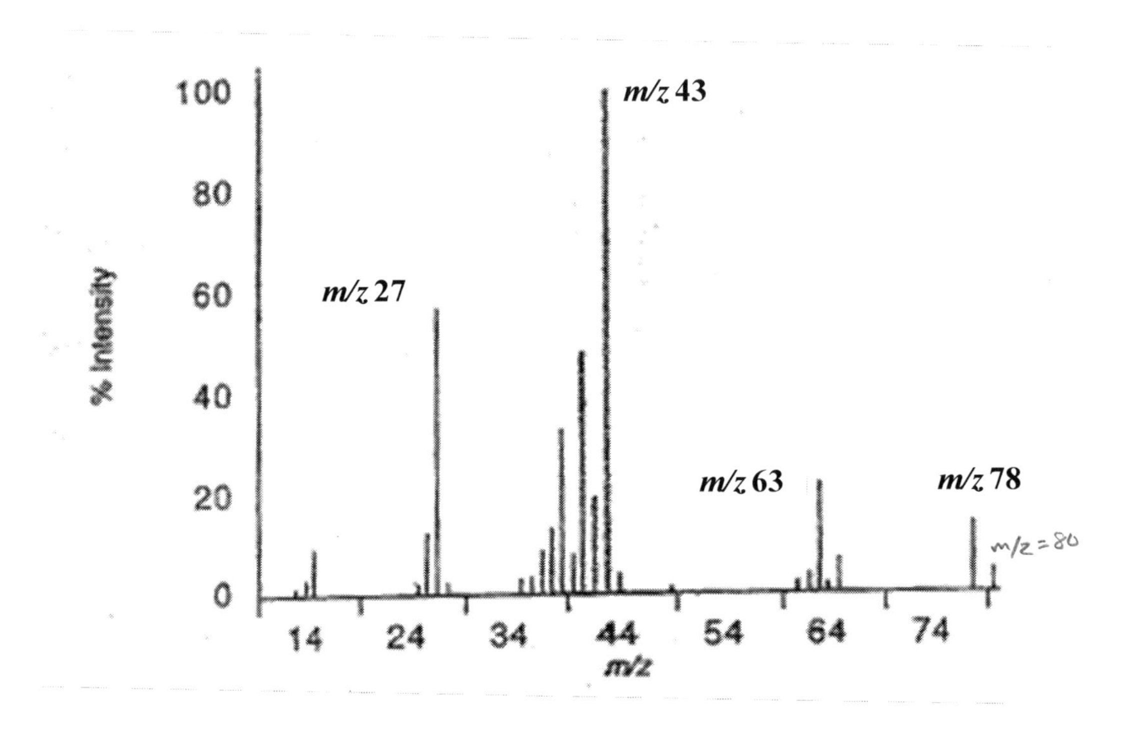

Notes:

- "M" & "M+2" must differ by 2 units if Cl is in comp.

An M + 2 peak that is 1/3 the height M of molecular ion:

Ⓐ Cl in Compound

The base peak (*m/z* = 43):

Ⓑ loss of Cl (size match)

***m/z* = 63 and *m/z* = 65:**

Ⓒ ↓ h ↓ 1/3 h (Not explained in class)

Notes:

Other Fragments?: α Cleaveage

Some functional groups can undergo more complex fragmentation via breakage of a bond alpha to the heteroatom substituent:

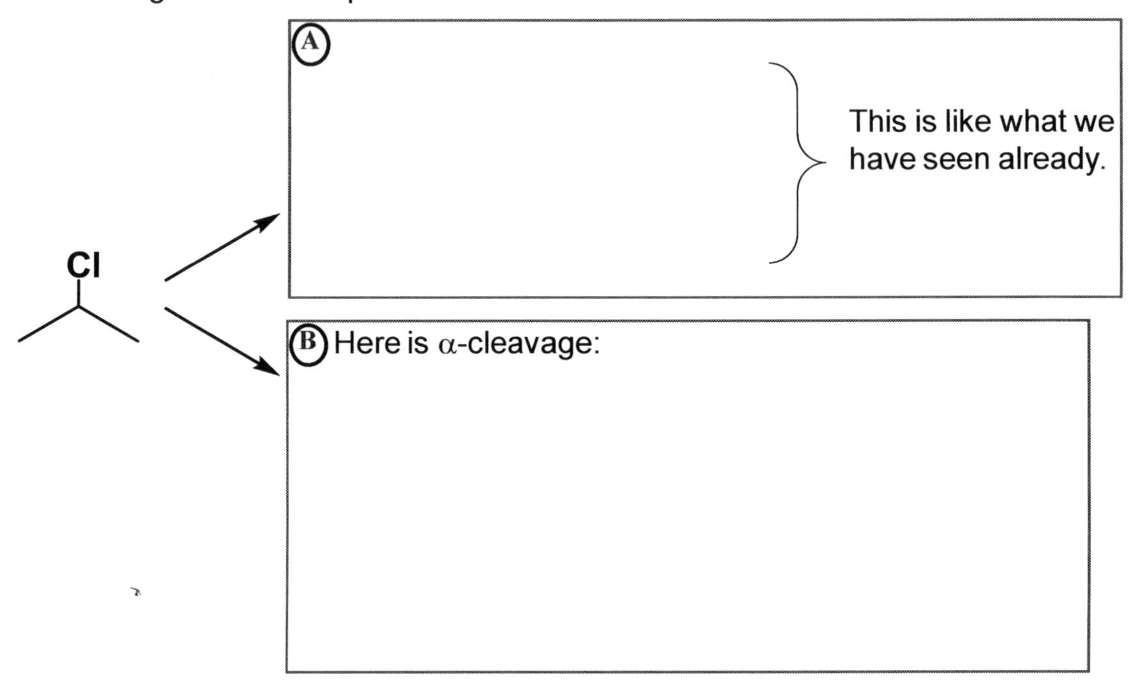

Notes:

s-Bu-O-^{i}Pr

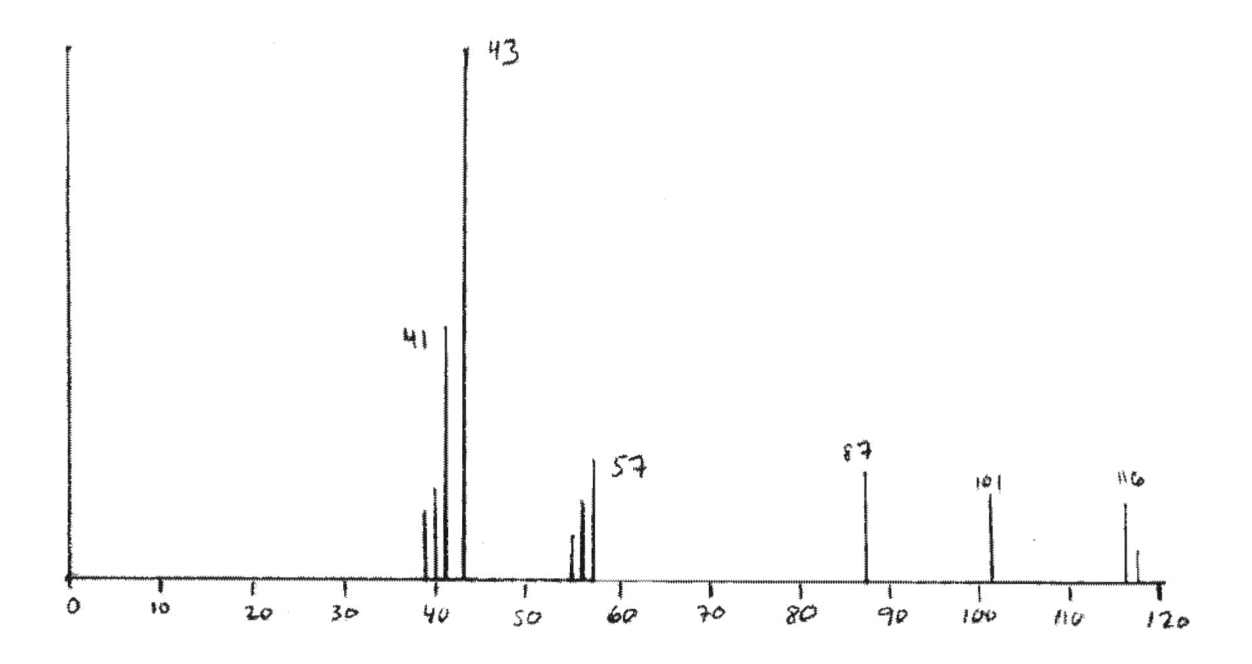

Notes:

Identification of the 116, 57, and 43 peaks is fairly straightforward:

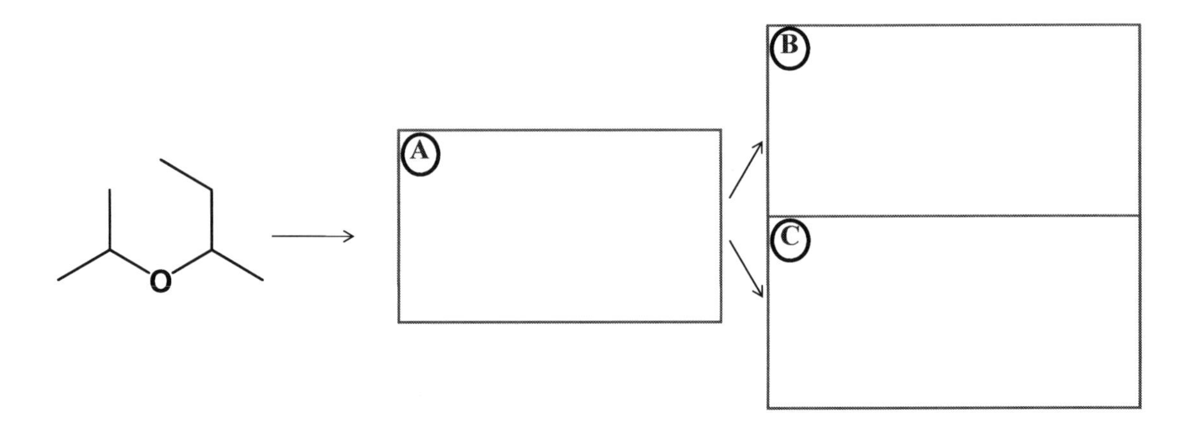

Other peaks can result from α cleavage ...

Notes:

Now consider some possible α cleavage products:

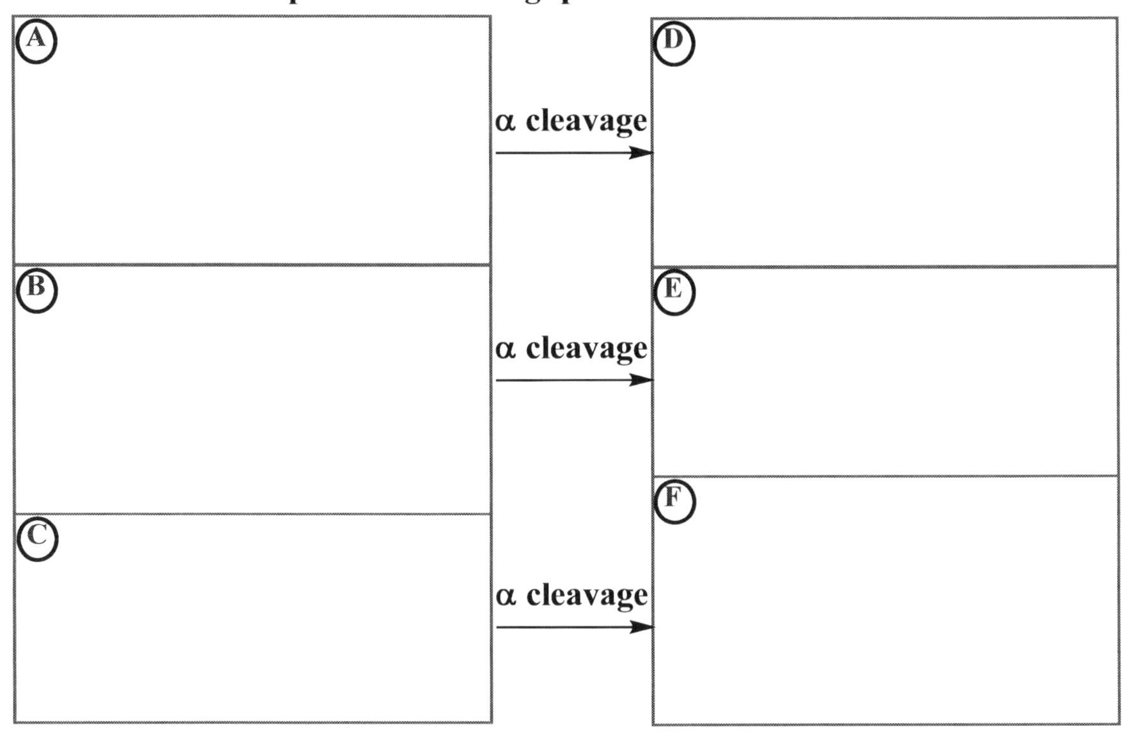

Notes:

2-Hexanol

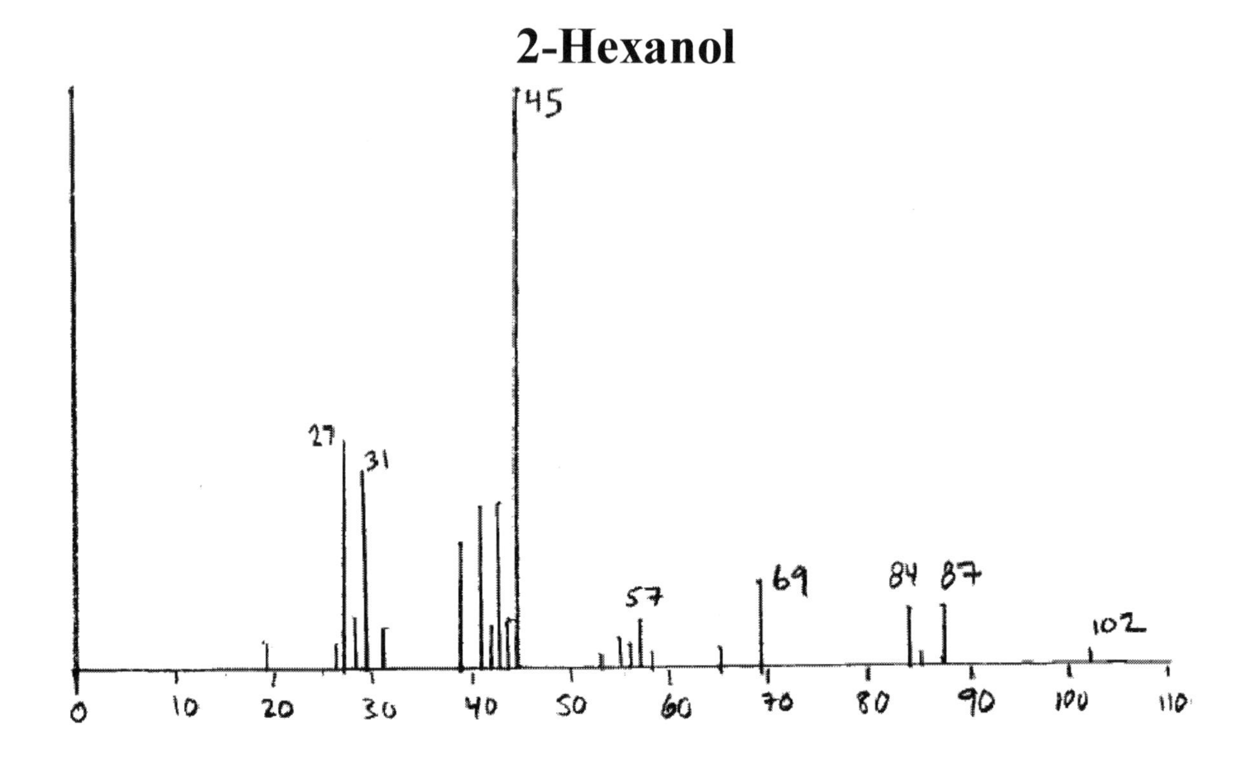

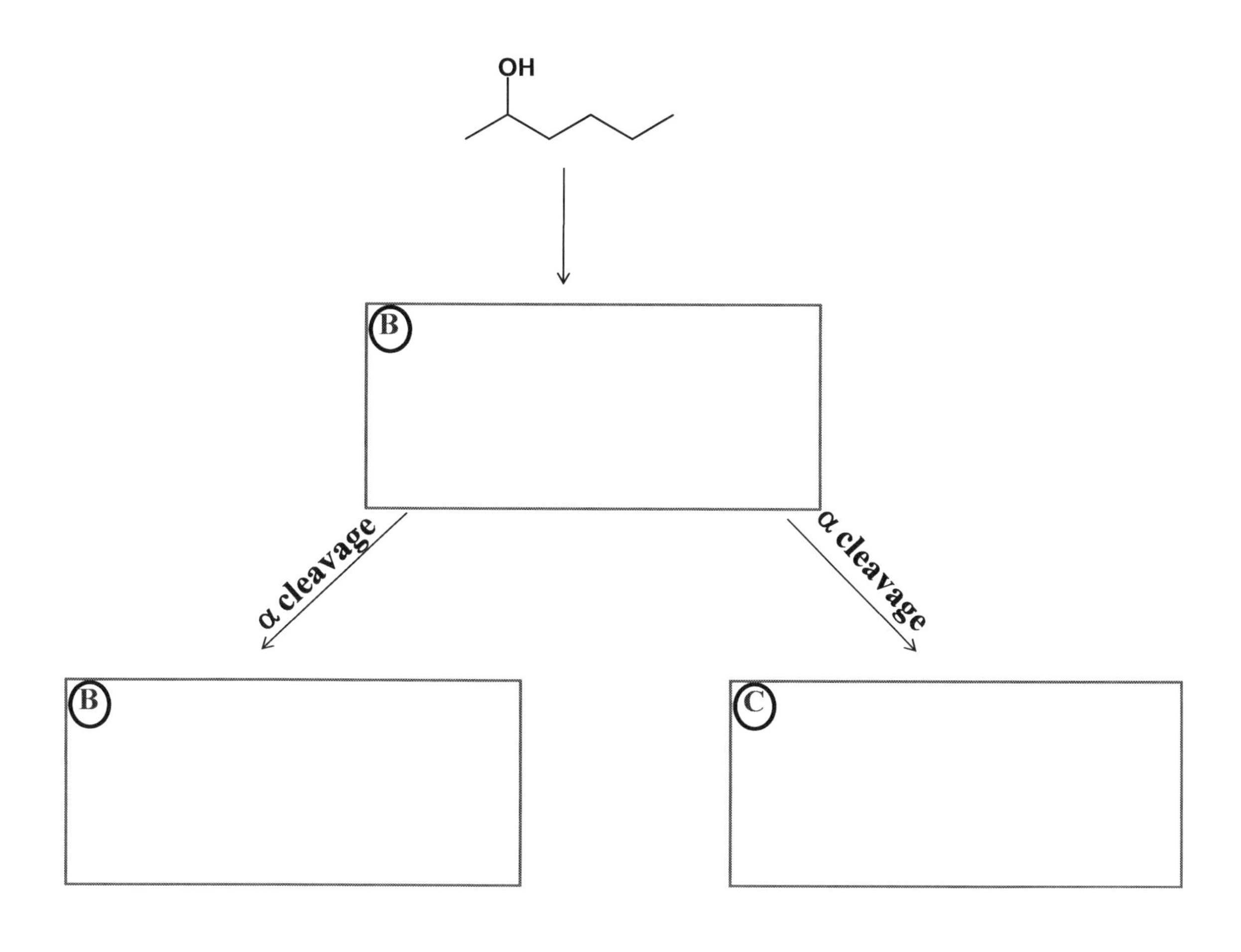
OH
B
α cleavage
α cleavage
B
C

Notes:

A special type of fragmentation is operative in alcohols in which water is eliminated by OH leaving with a γ-hydrogen:

H
O
H
α
β
γ
→
(A)

Notes:

Recap

(A)
(B)
(C)

Notes:

Fragmentation Pattern of Ketones

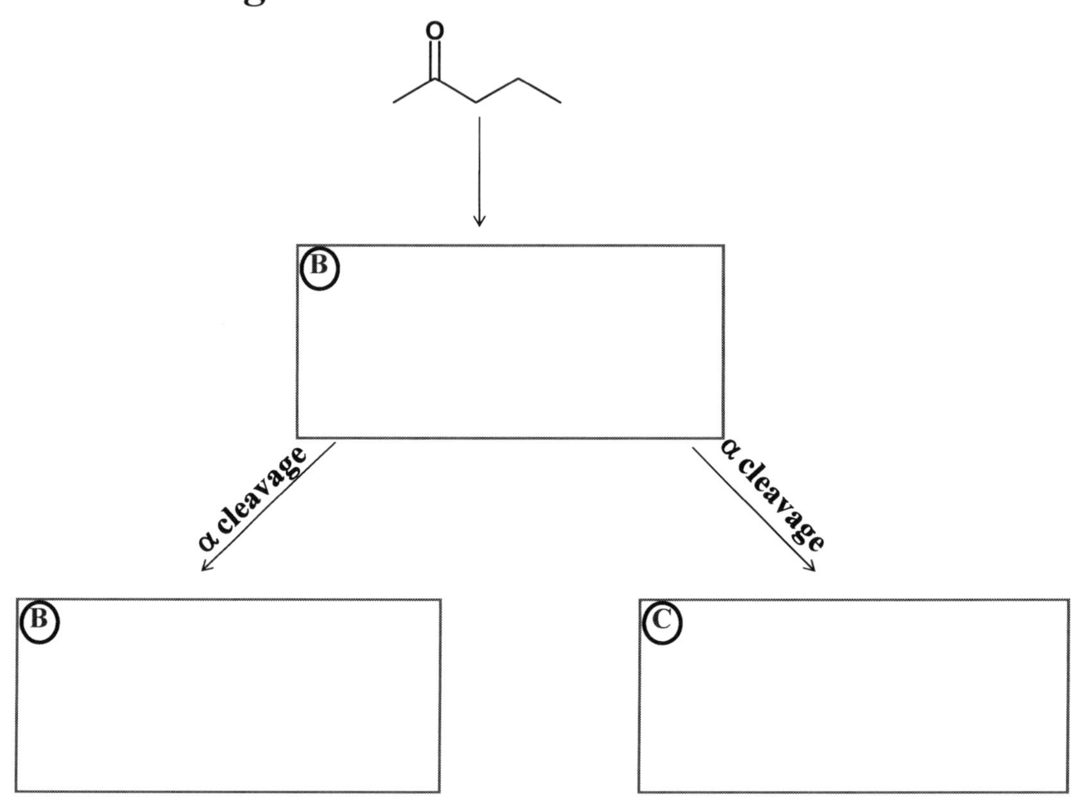

Notes:

McLafferty Rearrangement

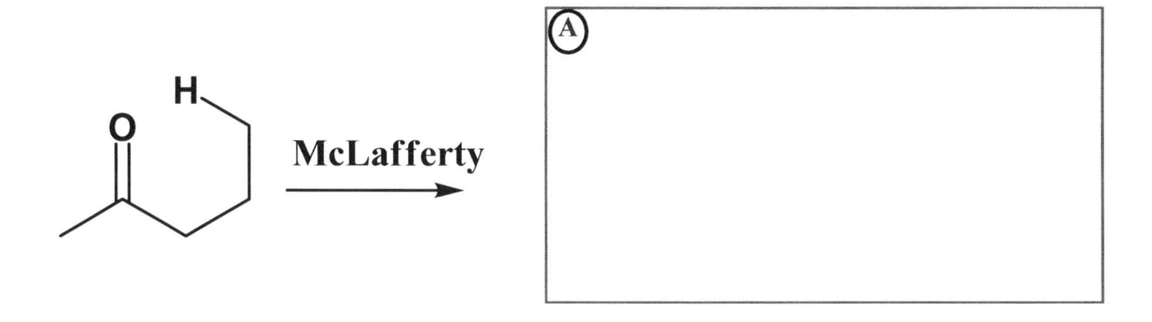

Notes:

Technique 2: Infrared Spectroscopy

Molecules absorb energy:

Ⓐ IR light shine on sample → different molecules absorb different wave lengths of IR light

Leading to:

Ⓑ vibrations, stretching, bending, etc.

Expressing energy: $E = h\upsilon = \frac{hc}{\lambda}$

↖ Planck constant

wavenumber ($\tilde{\upsilon}$):

Ⓒ $E \propto \frac{1}{\lambda}$ $\tilde{\upsilon}$ = a way to express E

Large wavenumbers =

Ⓓ correspond to higher energy

Notes:

Bonds can vibrate in various ways:

	Symmetric		asymmetric
Stretch			
bends		(in plane)	R
			away
			toward

A Each of these motions can have its own peak

Notes:

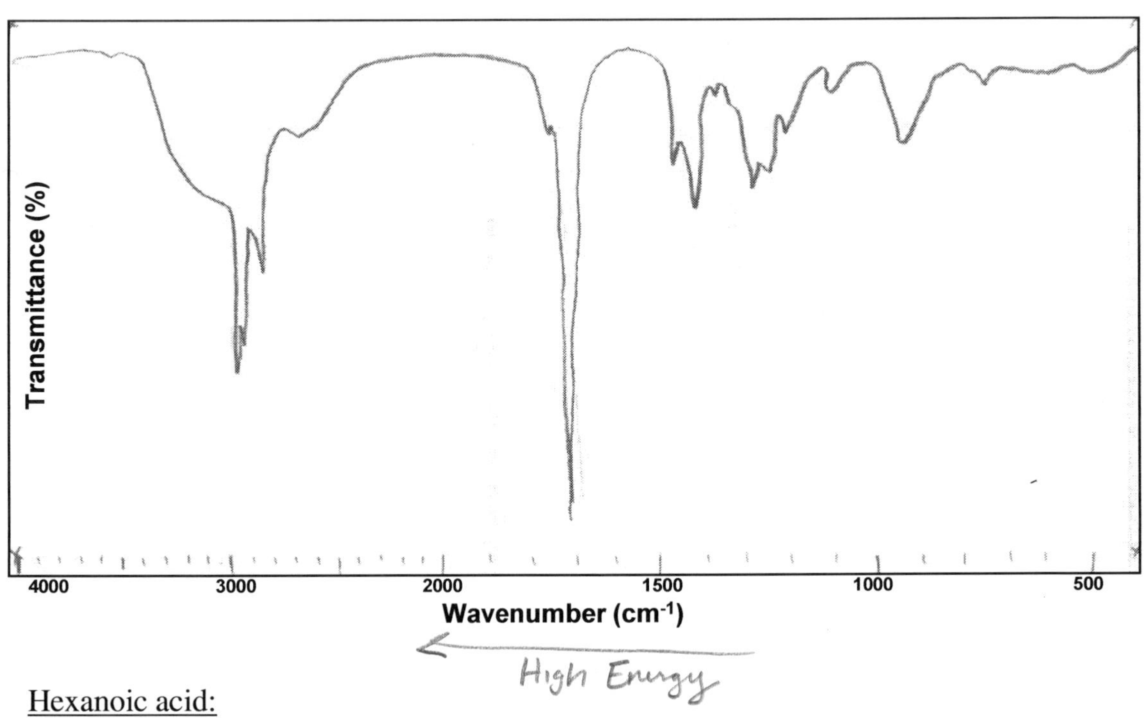

Hexanoic acid:

Notes:

- Sharp spikes mean absorbtion of Energy (1700 cm^{-1}) (bond w/ strength)

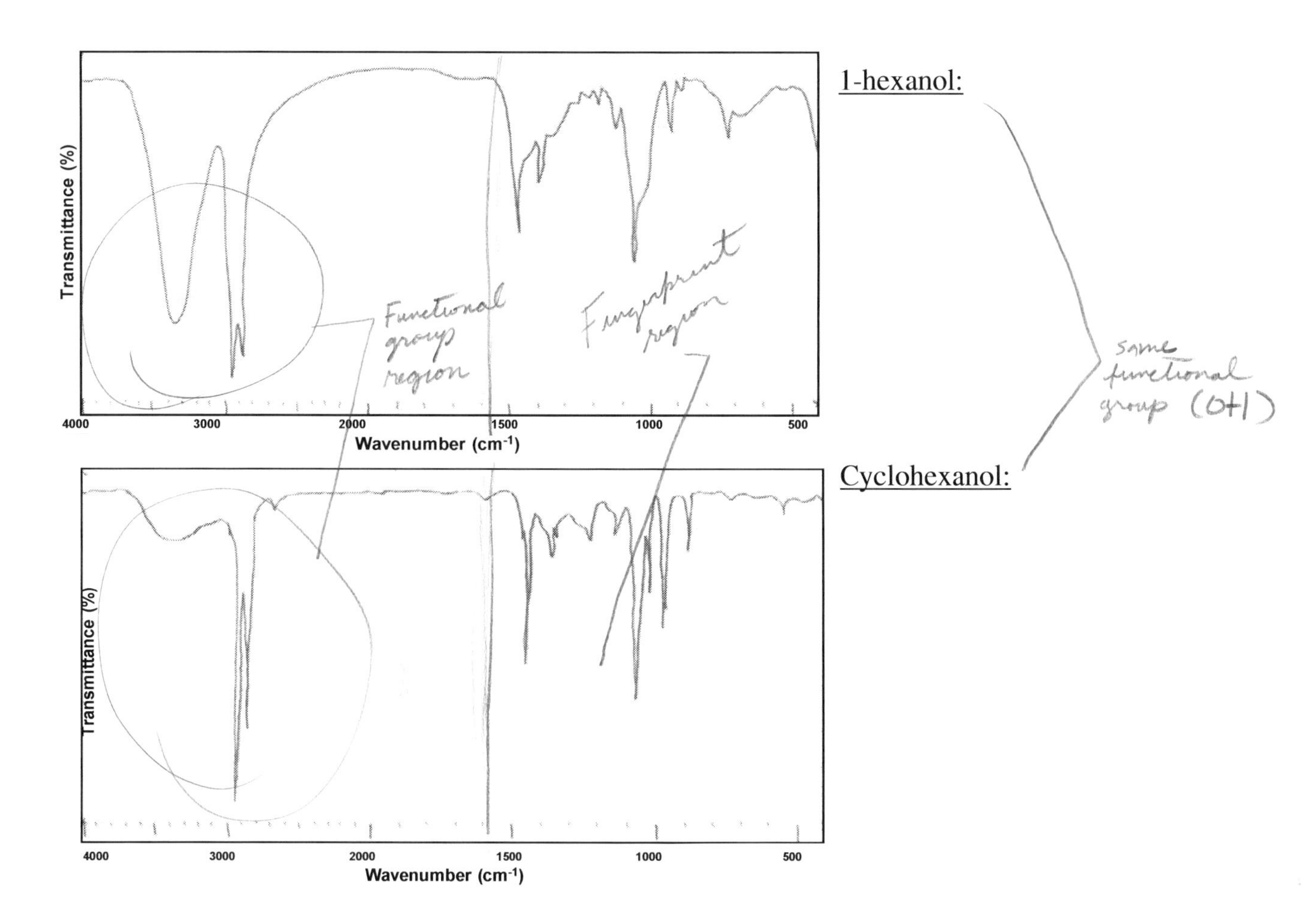
Transmittance (%)
4000
3000
2000
1500
1000
500
Wavenumber (cm-1)
1-hexanol:
Functional group region
Fingerprint region
same functional group (OH)
Cyclohexanol:

Notes:

Characteristic Energies (Stretches)

	Bond	Energy (cm^{-1})	Intensity
	N≡C	2255-2220	m
	C≡C	2260-2100	w-m
easier to stretch then C≡C →	C=C	1675-1660	m
	N=C	1650-1550	m
	benzene ring	1600 **AND** 1500-1425	w-s
	C=O	1775-1650	s
	C—O	1250-1000	s
	C—N	1230-1000	m
alcohol	O—H	3650-3200	s (br)
C(O)OH	O—H	3300-2500	s (br)
	N—H	3500-3300	m (br)
	C—H	3300-2725	m

more polar = harder to stretch (remember E.N.)

Notes:

Peak intensity depends on several factors:

Ⓐ Greater Dipole (more polar bond)
→ higher intensity

Ⓑ stretches increase dipole

Ⓒ more bonds = more intense

Notes:

Bond strength is effected by neighbors:

(A)

1. e⁻ delocalization (resonance)
2. e⁻ w/d or donating groups (inductive effect)
3. H-bonding

Consider conjugated vs nonconjugated C=O:

1720 cm^{-1}

non-conj.

vs.

1680 cm^{-1}

conjugated

Notes:

The C=O stretch depends on which carbonyl functional group is in the compound:

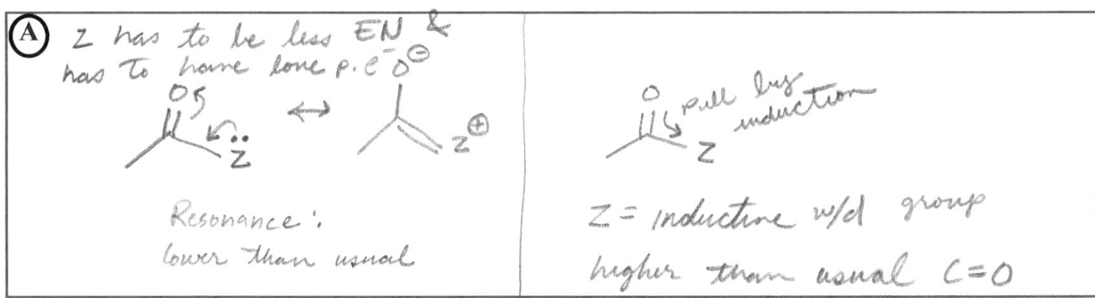

A
Z has to be less EN &
has to have lone p.e⁻
Resonance:
lower than usual
pull by induction
Z = inductive w/d group
higher than usual C=O

Amide:

B → resonance so C=O stretch less than usual
less EN than "O"

Ester:

C → equivalent EN → Inductive effect → C=O higher than Ketone

Notes:

Consider how IR can differentiate groups with different types of C-O bonds

Alcohol

(A) C-O at 1050 cm^{-1}
O-H at 3200 cm^{-1}

Ether

(B) C-O at 1050 cm^{-1}

Carboxylic acid

(C) O-H 3200 cm^{-1}
C=O 1700 cm^{-1}

Ester

(D) C=O ~1720 cm^{-1}
no O-H bond

Notes:

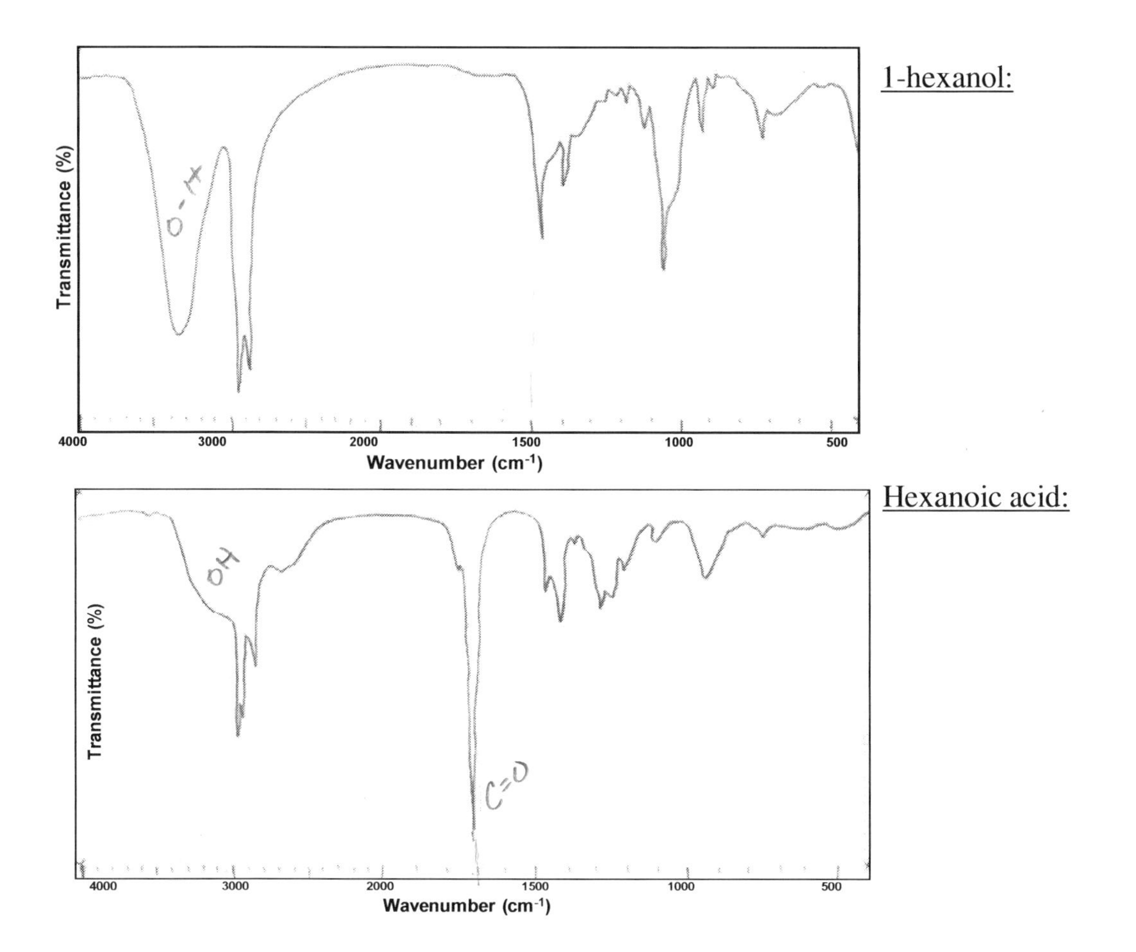
1-hexanol:
Transmittance (%)
O-H
4000
3000
2000
1500
1000
500
Wavenumber (cm-1)
Hexanoic acid:
Transmittance (%)
OH
C=O
4000
3000
2000
1500
1000
500
Wavenumber (cm-1)

Notes:

Factors influencing O–H absorption:

(A)

Energy of absorption:

(B)

concentrated

(C)

dilute

(D)

Notes:

C-H Stretches:

Isomerism:

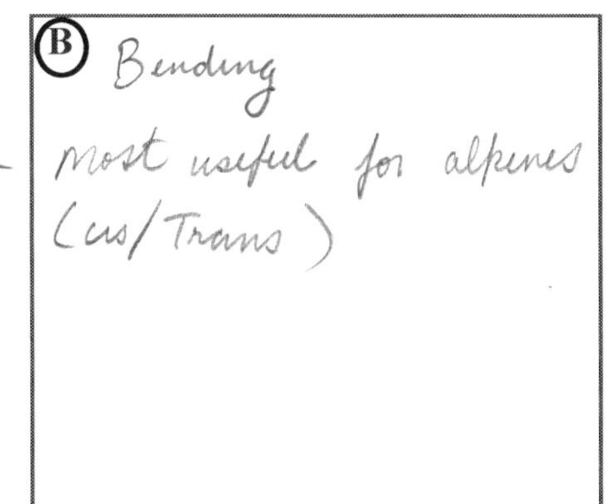

C-H Bond (Stretch)	Energy (cm^{-1})
C≡C—H (sp)	3300-ish
C=C—H (sp^2)	3100-3000
C—C—H (sp^3)	2950-2850
Aldehyde C(=O)—H	2820-ish and 2720-ish

C-H Bond (Bending)

Group	Energy (cm^{-1})
—CH_3, —CH_2—, —CH(H)—	1450-1400
RHC=CHR (R and H trans)	980-960 trans
RHC=CHR (R, R on same side)	730-670 cis
R_2C=CHR	840-800 trisubstituted
RHC=CH_2	990 and 910 monosubstituted
R_2C=CH_2	890 disubstituted terminal

Notes:

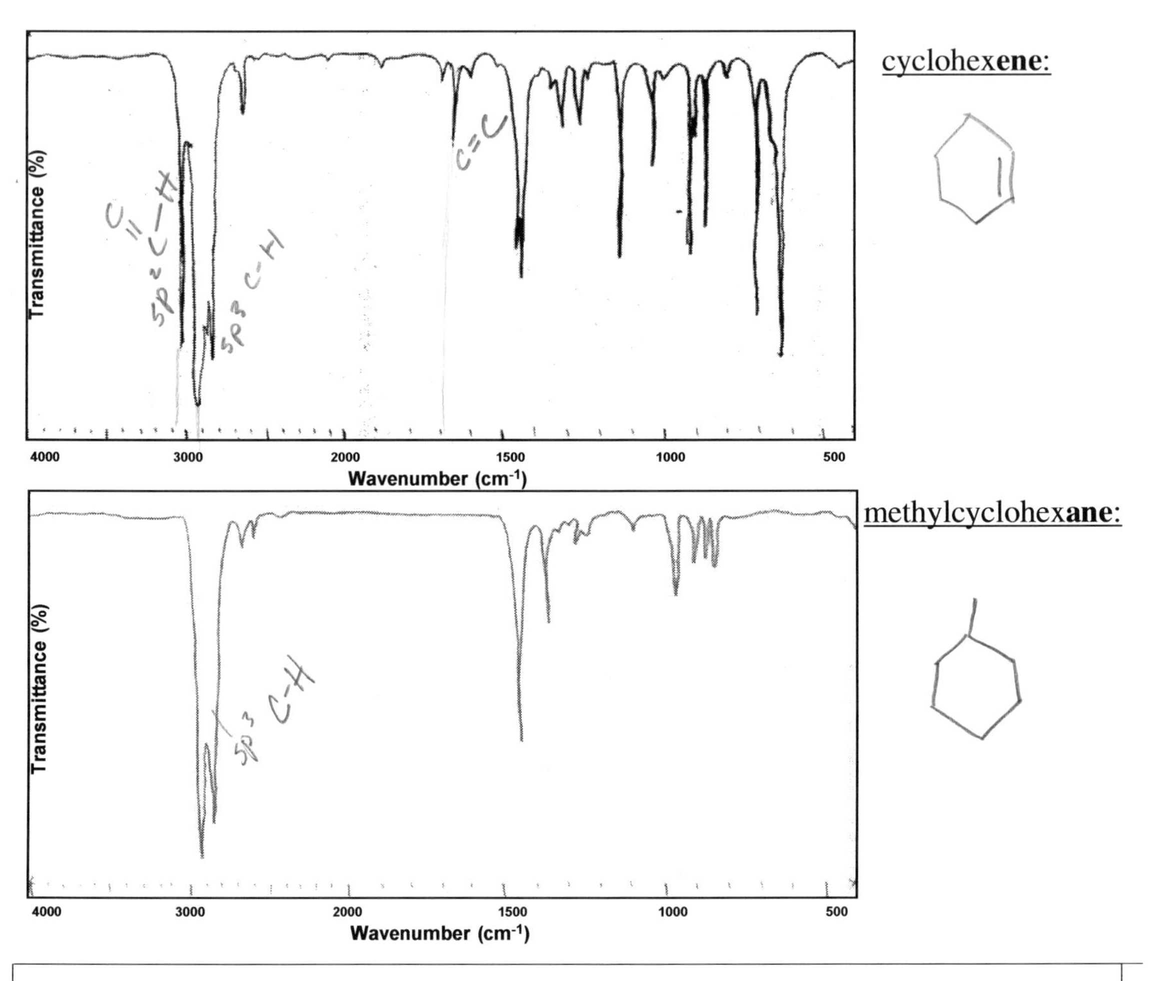

cyclohexene:
Transmittance (%)
sp² C—H
sp³ C-H
C=C
4000
3000
2000
1500
1000
500
Wavenumber (cm⁻¹)
methylcyclohexane:
Transmittance (%)
sp³ C-H
4000
3000
2000
1500
1000
500
Wavenumber (cm⁻¹)

Notes:

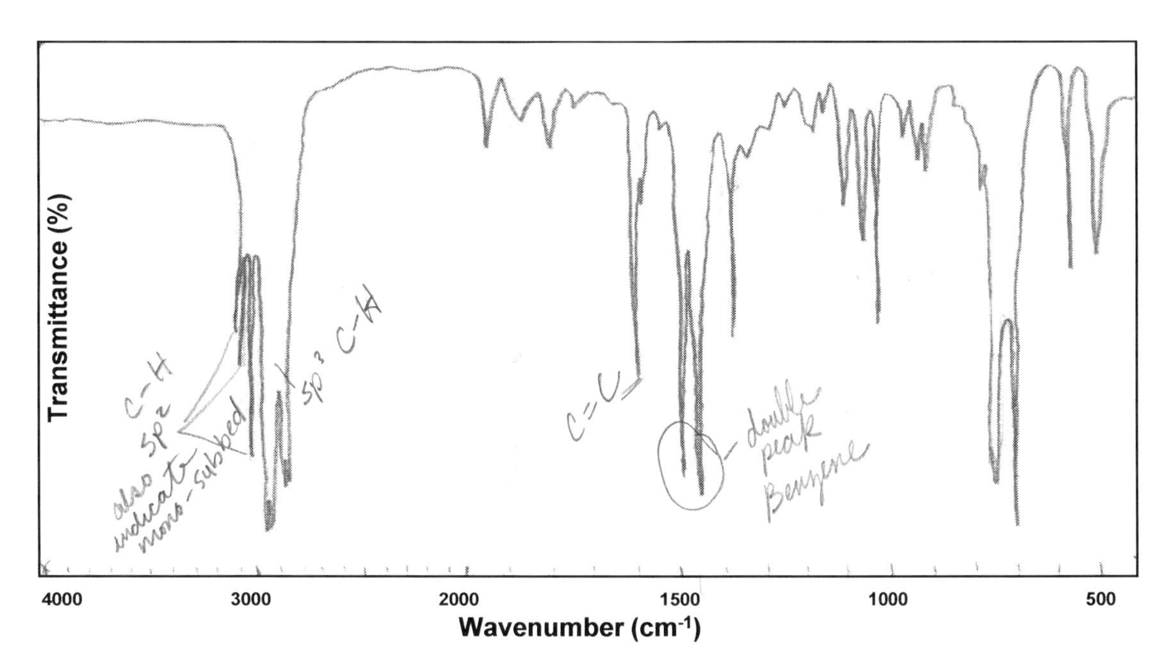

n-butylbenzene:

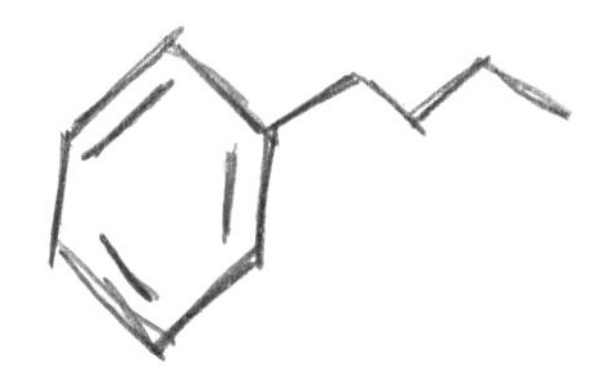

Notes:

N–H bending or C=C? at ~1600 cm^{-1}??

(A)

AND:

(B)

Notes:

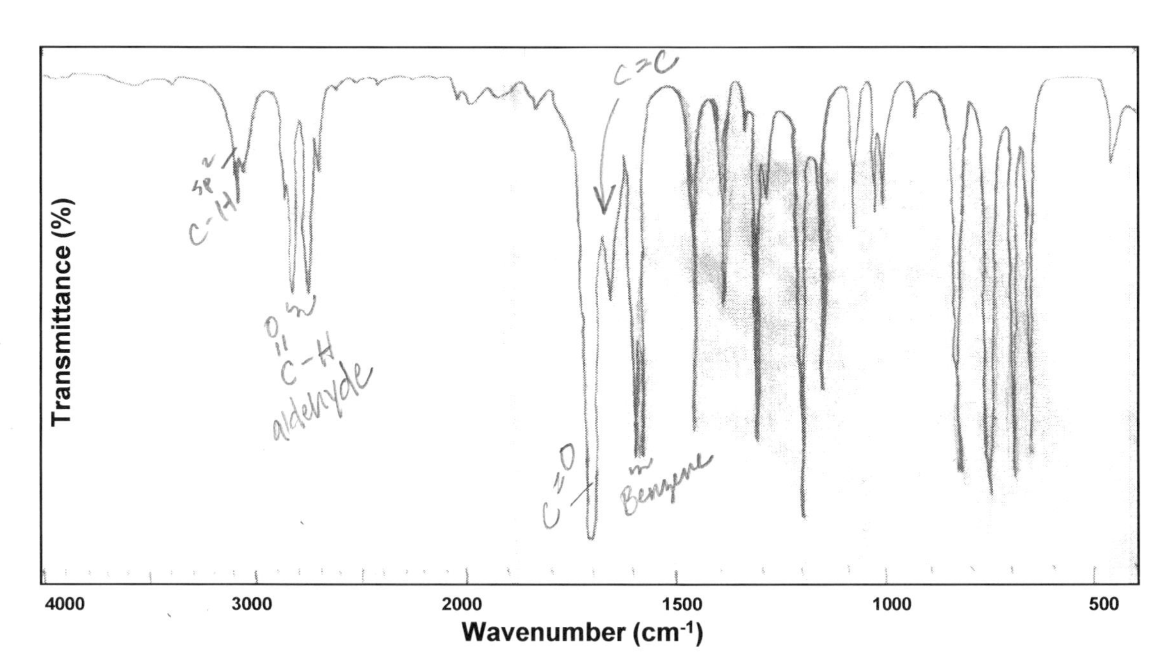

benzaldehyde:

Notes:

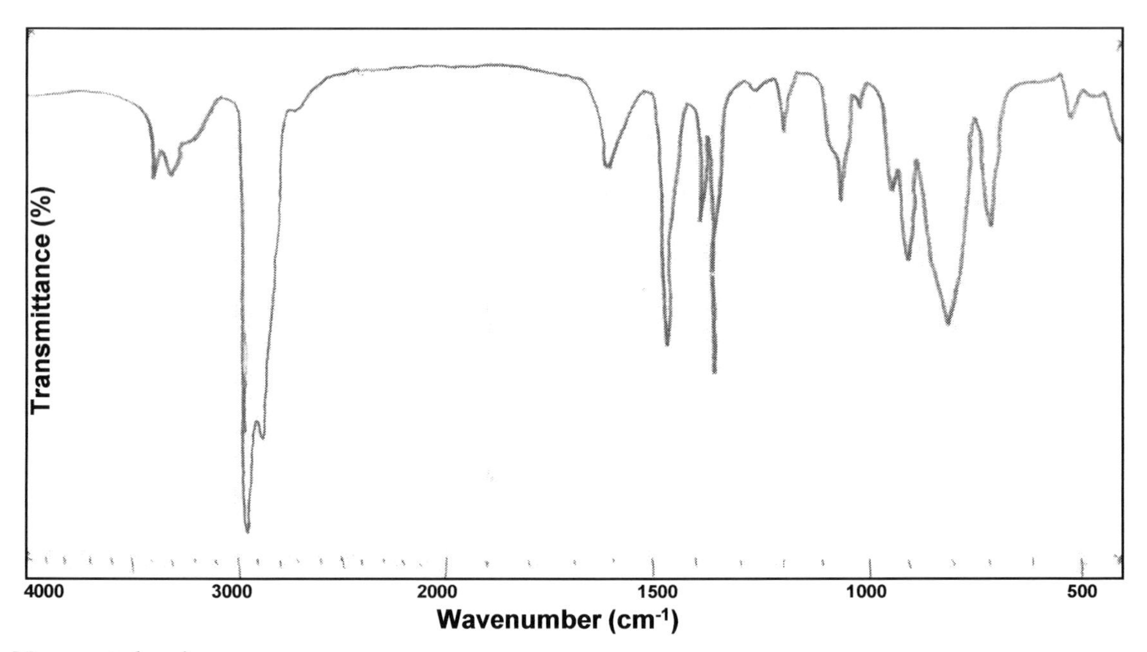

Neopentylamine:

Notes:

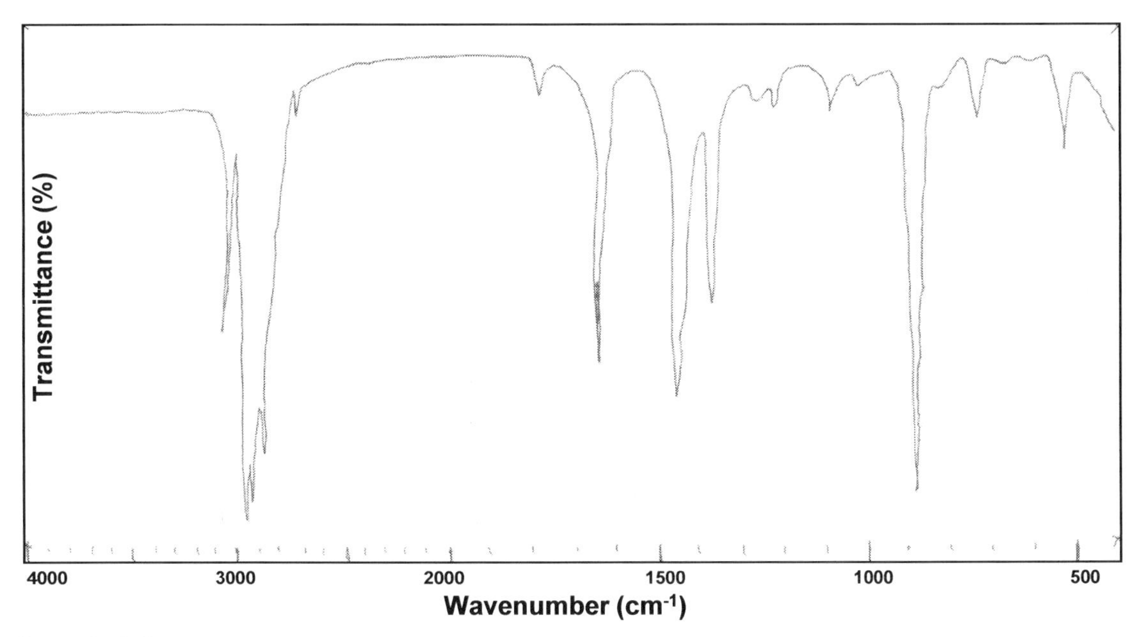

2-methyl-1-pentene:

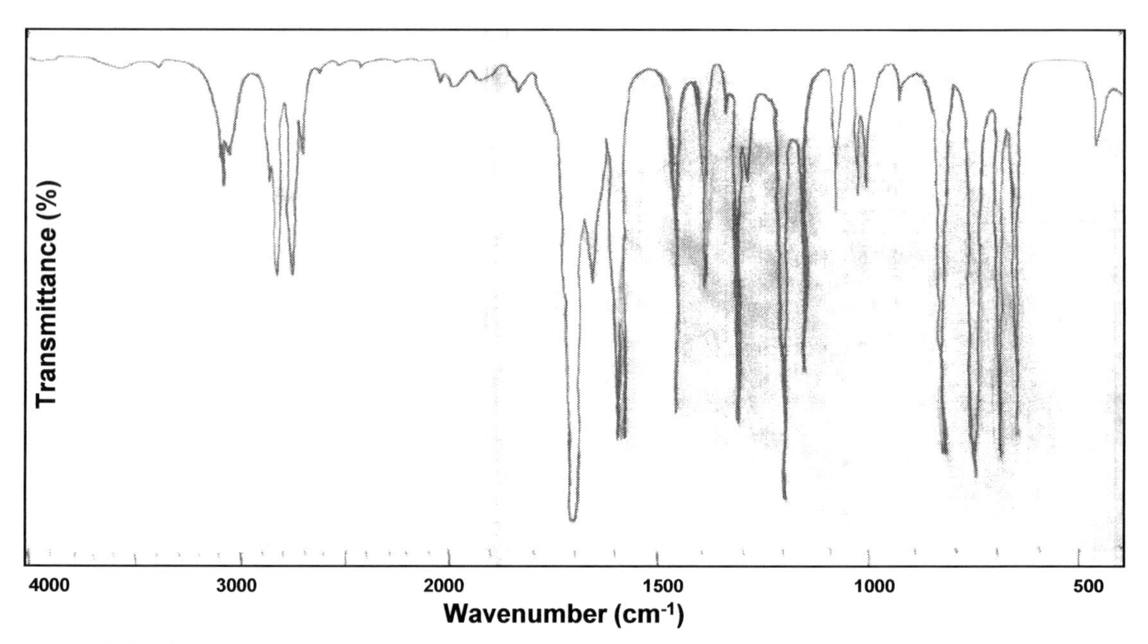

benzaldehyde:

Notes:

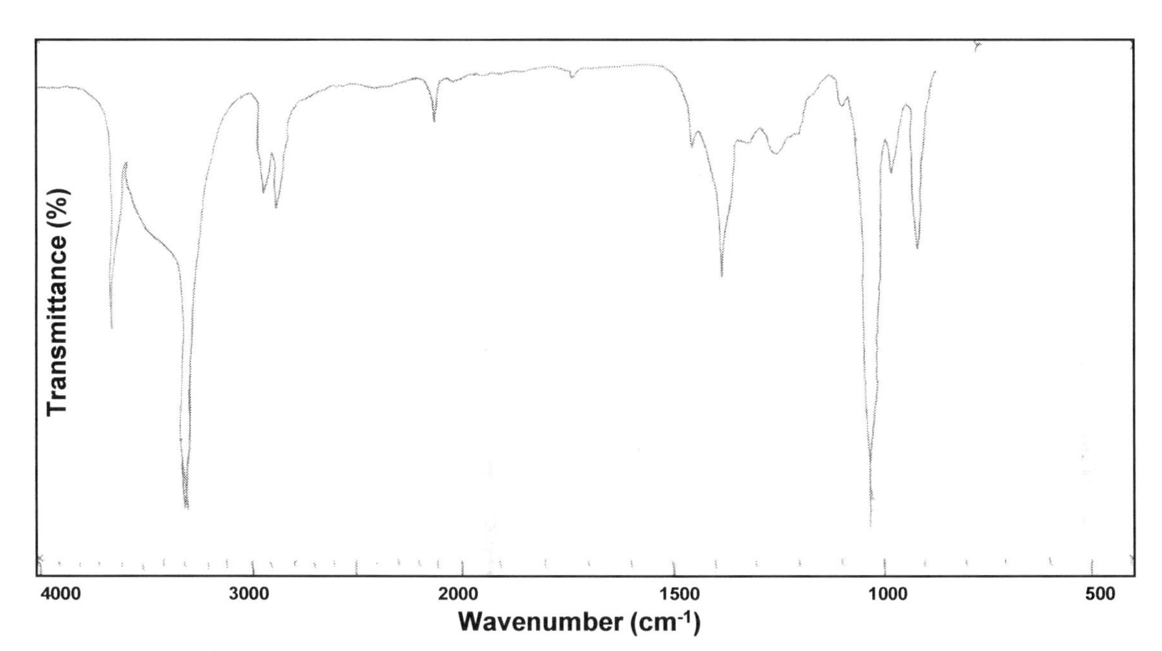

2-propyne-1-ol:

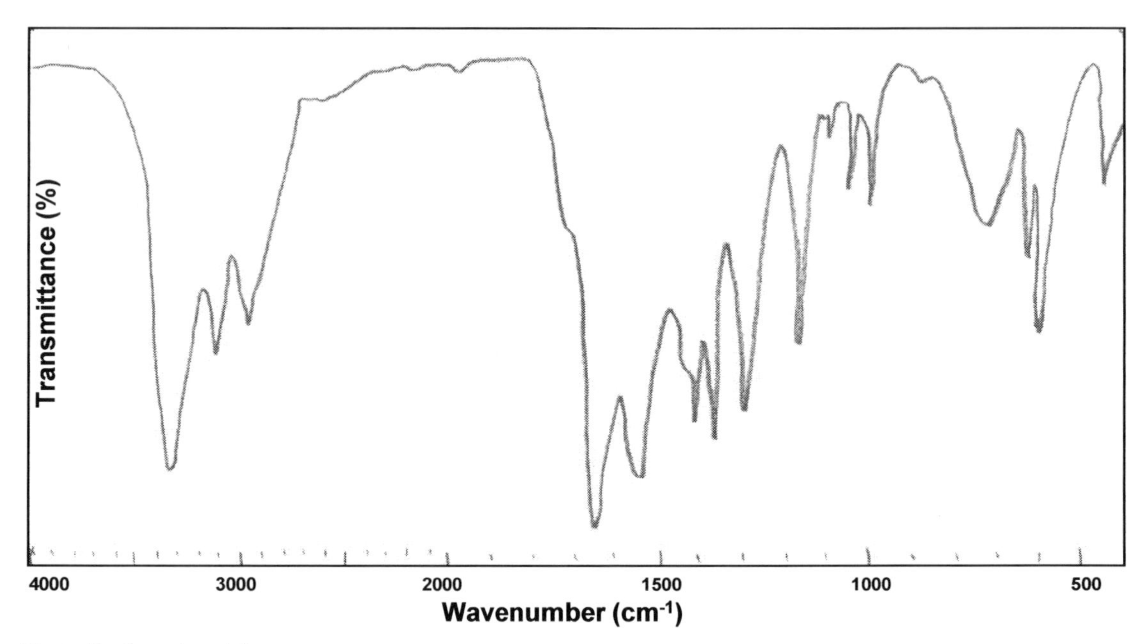

N-methylacetamide:

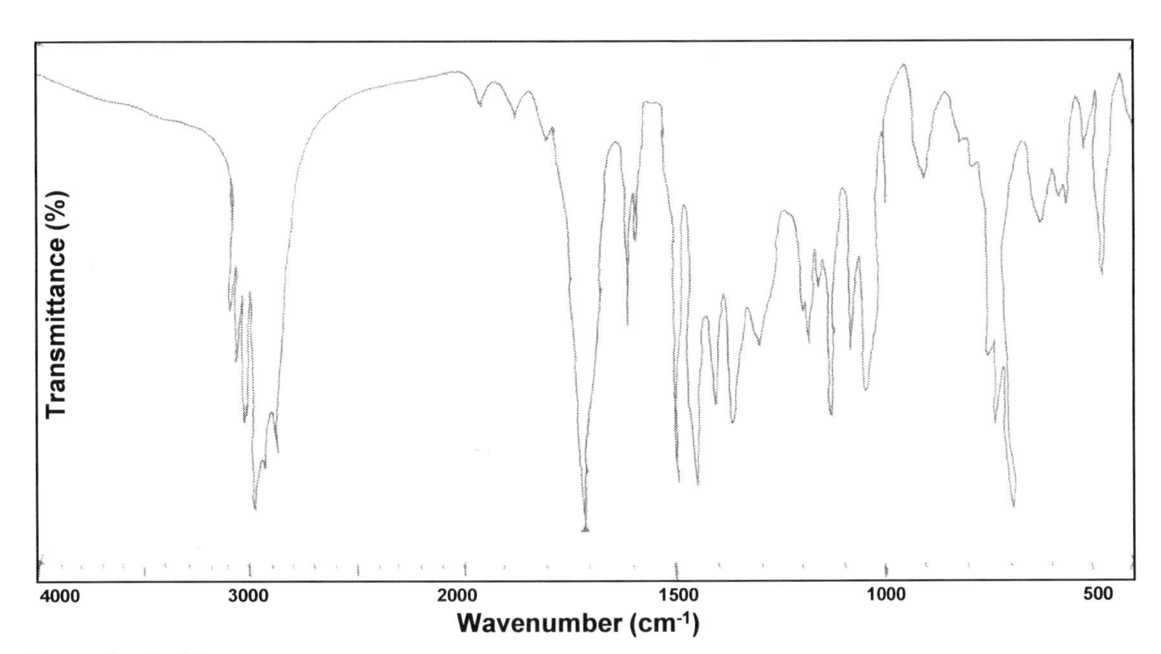

Benzyl ethyl ketone:

Lecture Set 3: Nuclear Magnetic Resonance Spectroscopy

Suggested Reading:

Suggested Problems Text:

What is NMR?

NMR stands for:

Ⓐ Nuclear Magnetic Resonance

NMR spectroscopy is a technique that is used to identify compounds.
A NMR Spectrum is a plot of

Ⓑ Energy

vs.

Ⓒ Intensity

A representative spectrum is shown here:

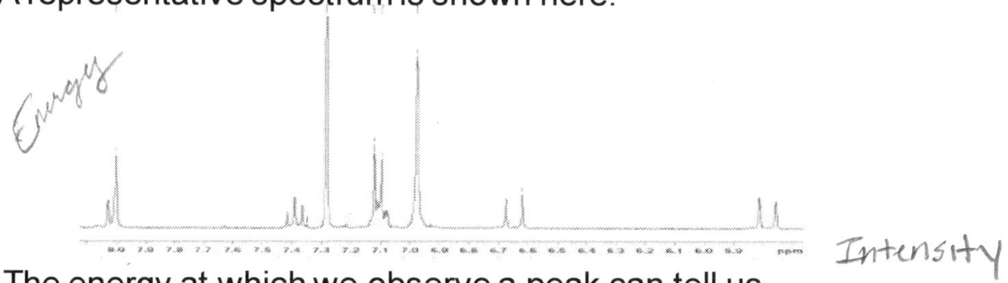

Energy

Intensity

The energy at which we observe a peak can tell us

Ⓓ e^- Density around nucleas

thus aiding in the compound's identification.

Notes:

How Does NMR Spectroscopy Work?

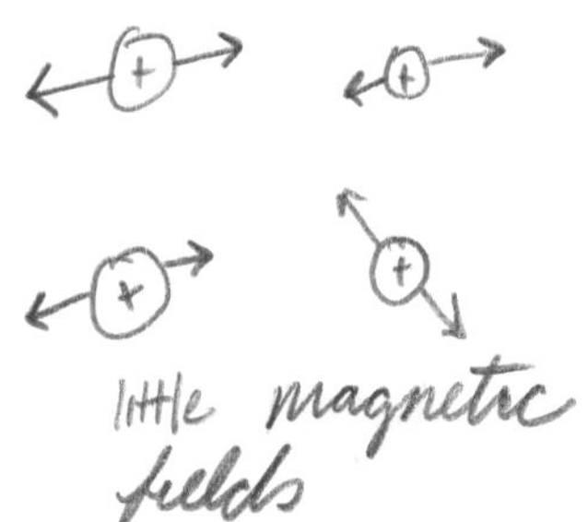

Nuclei are charged. Charged particles interact with magnetic fields. A 'resonating' nucleus generates a magnetic field of its own. This generated magnetic field may be aligned with or oppose the applied field. It takes more energy to oppose the applied field. One can add energy to get the nucleus' magnetic field to 'flip' direction. By measuring the energy needed to accomplish this flip, we generate a NMR spectrum.

A different magnetic field strength is needed for each atomic nucleus (i.e., it is different for ^{1}H versus ^{13}C versus ^{31}P, etc. So for simple NMR spectra, we only look at one type of nucleus at a time. In Organic Chemistry, we will focus on ^{13}C and ^{1}H NMR spectroscopy, because most organic molecules have a lot of C and H atoms.

Notes:

^{1}H NMR Lesson I:
Not all ^{1}H nuclei are equal!

Not all protons are the same, even those within this simple molecule. If the protons are different, they will 'flip' at different energies, giving us different peaks in the spectrum.

But what makes these ^{1}H nuclei different from one another, energetically?

THINK about what we know:

1. We are looking at energy events within a magnetic field.
2. The nuclei interact with the magnetic field because they are charged.

Now, we also know that:

The nuclei also have electrons around them in the molecule
and
electrons are charged as well, so they will also interact with a magnetic field.

What effect will electrons have on the energy of the nuclei flipping?

Notes:

Conformational Interconversion and Signal Averaging

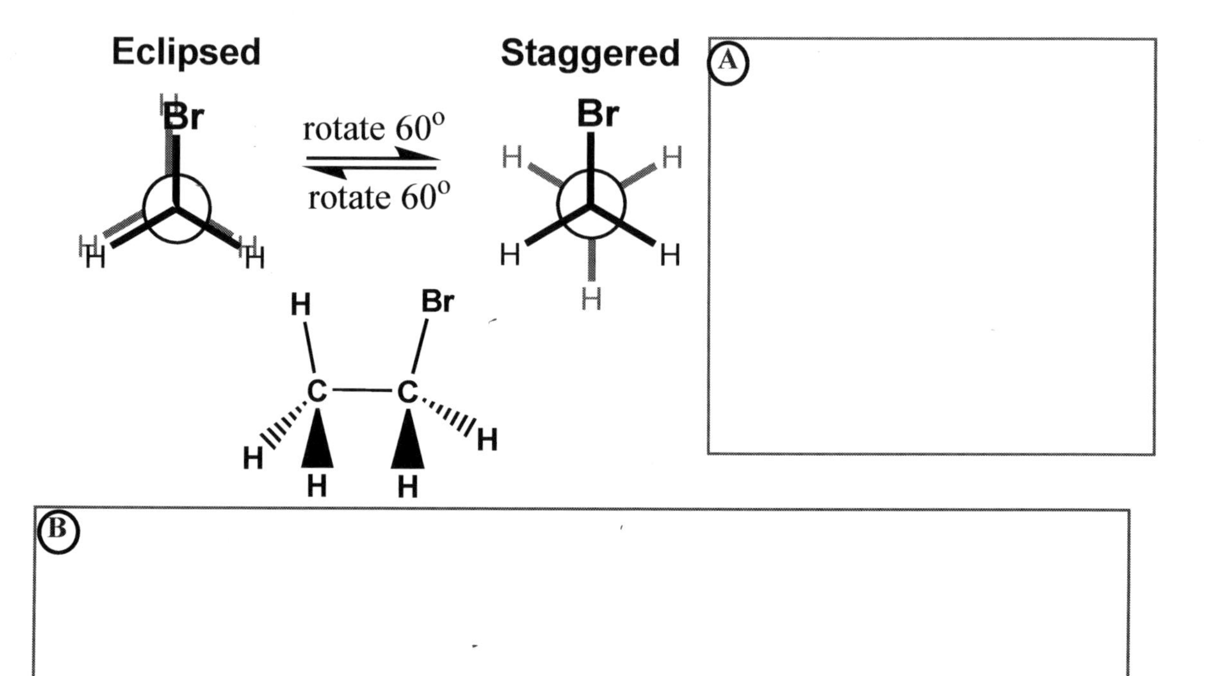

Notes:

* How Many Signals?

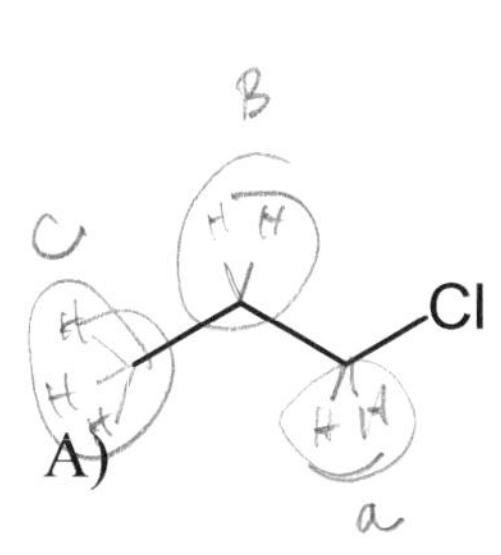

A)

3 signals b/c distance

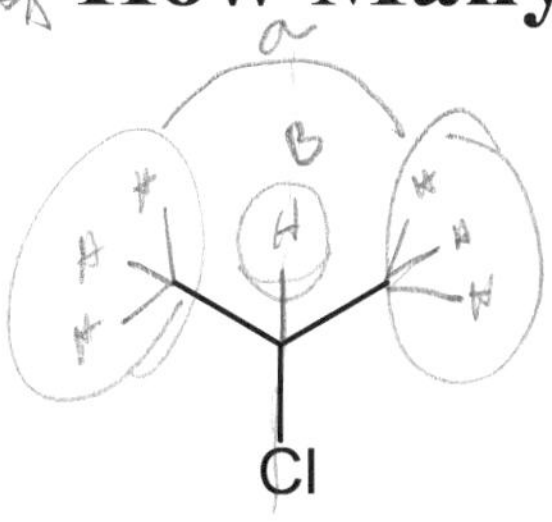

B)

2 signals b/c distance

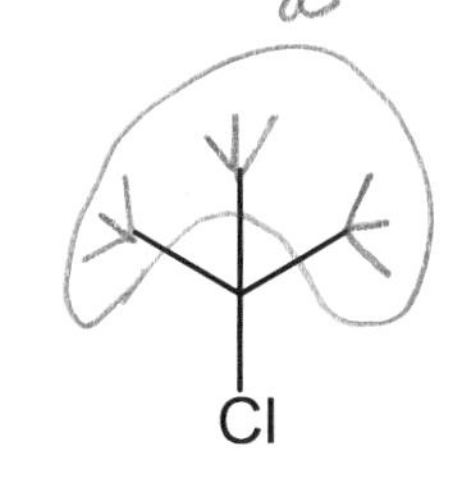

C)

1 signal

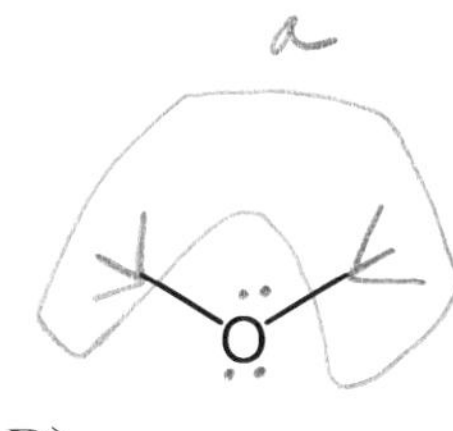

D)

1 signal

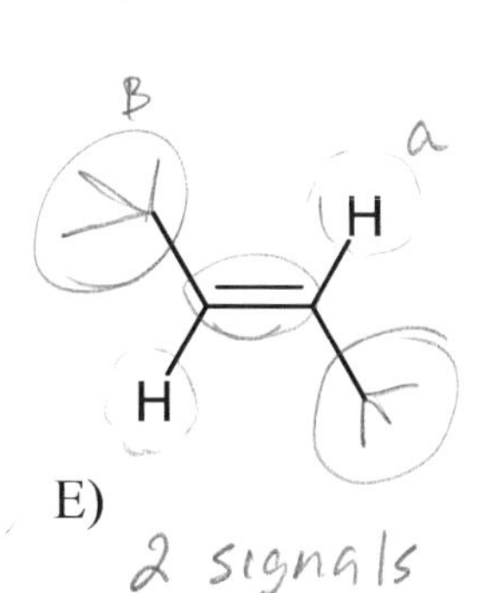

E)

2 signals

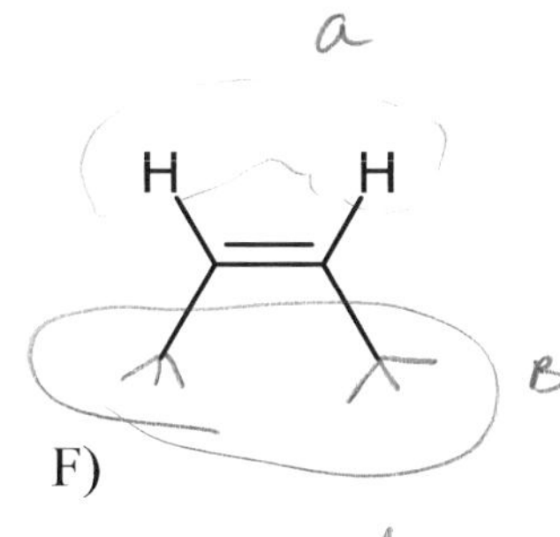

F)

2 signals

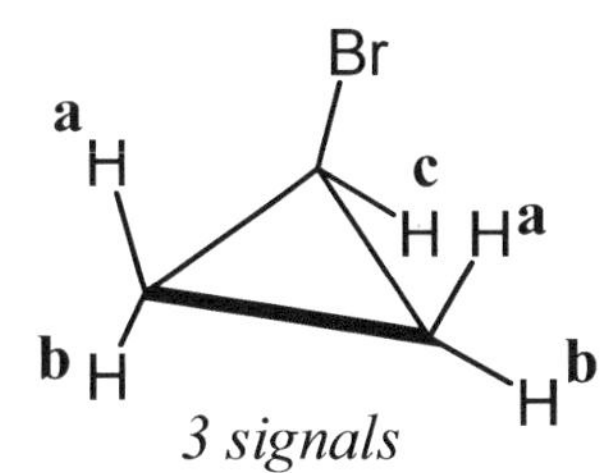

3 signals
3 magnetically inequivalent proton sets

*Always do Rings in 3-Dimensions *

Notes:

- for (E.) you can flip molecule or rotate 180° if molecules remain the same then NMR gives same reading

SKIP

Electrons Density Variation

Luckily, we've been evaluating variations in electron density since general chemistry:

Notes:

Electrons are Shields

$\vec{E}$ $\vec{f}$ $\vec{E}$ $\vec{e^-}$

$\vec{E} - \vec{e^-} = \vec{f}$

more e^- density around nucleus → less magnetic field
→ lower energy for that peak

Notes:

Inductive Effect

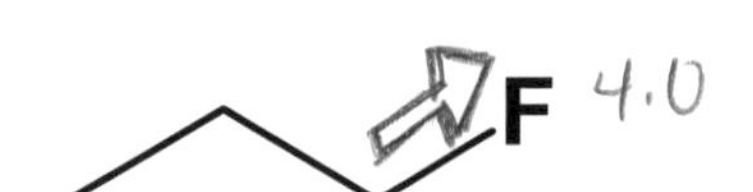

Ⓐ

EN: 4.0

δ: 4.5

Cl 3.0

Ⓑ

EN: 3.0

δ: 3.5

Ⓒ

EN: 2.8

δ: 3.4

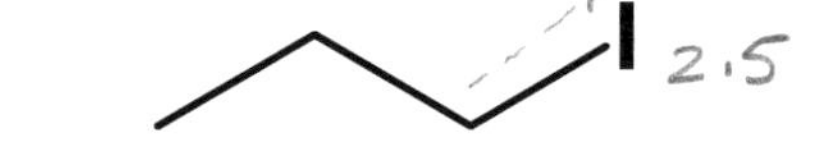

Ⓓ

EN: 2.5

δ: 3.2

Notes:

Terminology and Trends

Protons in Electron Poor environment:

(A) Deshielded
Downfield

↑ Energy

Protons in Electron Rich environment:

(B) Shielded
Upfield

↓ Energy

←

Frequency increases ν

δ Increases

Higher number on x axis in spectrum

Notes:

Trends in Chemical Shift

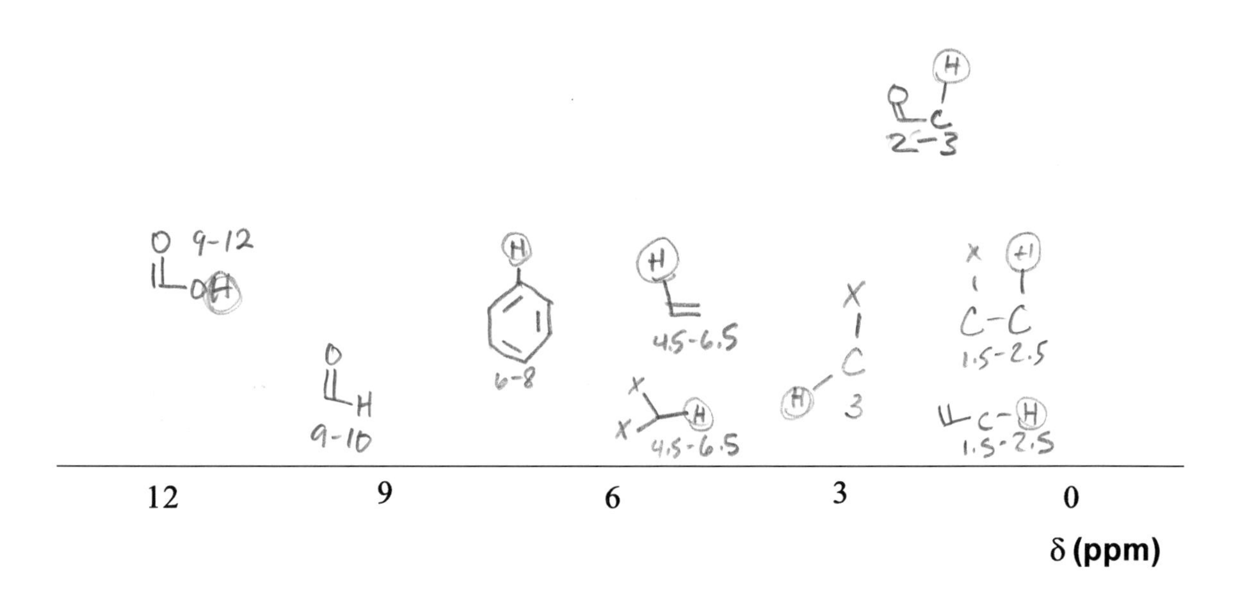

Notes:

Protons (Shown)	**Chemical Shift**
$Si(CH_3)_4$	0
$—CH_3$	0.9
$—CH_2—$	1.2
$—\overset{H}{C}—$	1.4
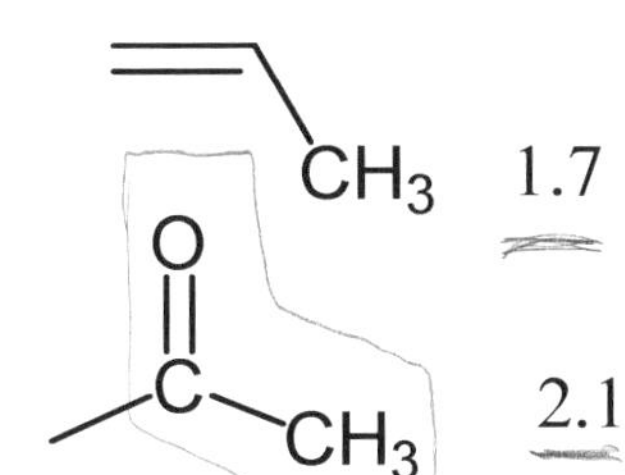 CH_3	1.7
CH_3	2.1
$C_6H_5—CH_3$	2.3
$≡—H$	2.4

Protons (Shown)	**Chemical Shift**
$R—OCH_3$	3.3
$=\!\!\diagdown H$	4.5-5.5
H—C—X, X = I	2.5-4
X = Br	2.5-4
X = Cl	3-4
X = F	4-4.5
$C_6H_5—CH_3$	6.5-8.0
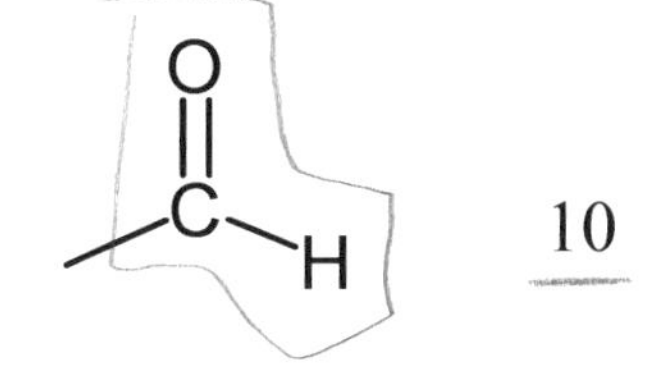 H	10

Notes:

^{1}H NMR Lesson II:
^{1}H nuclei know their neighbors!

If there is an NMR active nucleus (i.e., one which resonates thus creating a magnetic field) near another NMR-active nucleus, the two will influence each other. HOW? Let's investigate …

Notes:

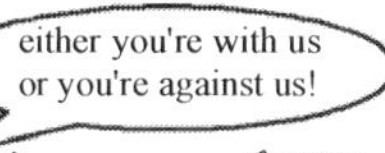

^{1}H NMR Lesson II: ^{1}H nuclei know their neighbors!

Let's say you have one neighbor. You can work with your neighbor or you can oppose your neighbor's effort to oppose the force of the boulder.

Notes:

Mr. Resonance's Neighborhood

Now you have two neighbors. This leads to three possible energy states for you as you attempt to hoist the huge-gantic boulder:

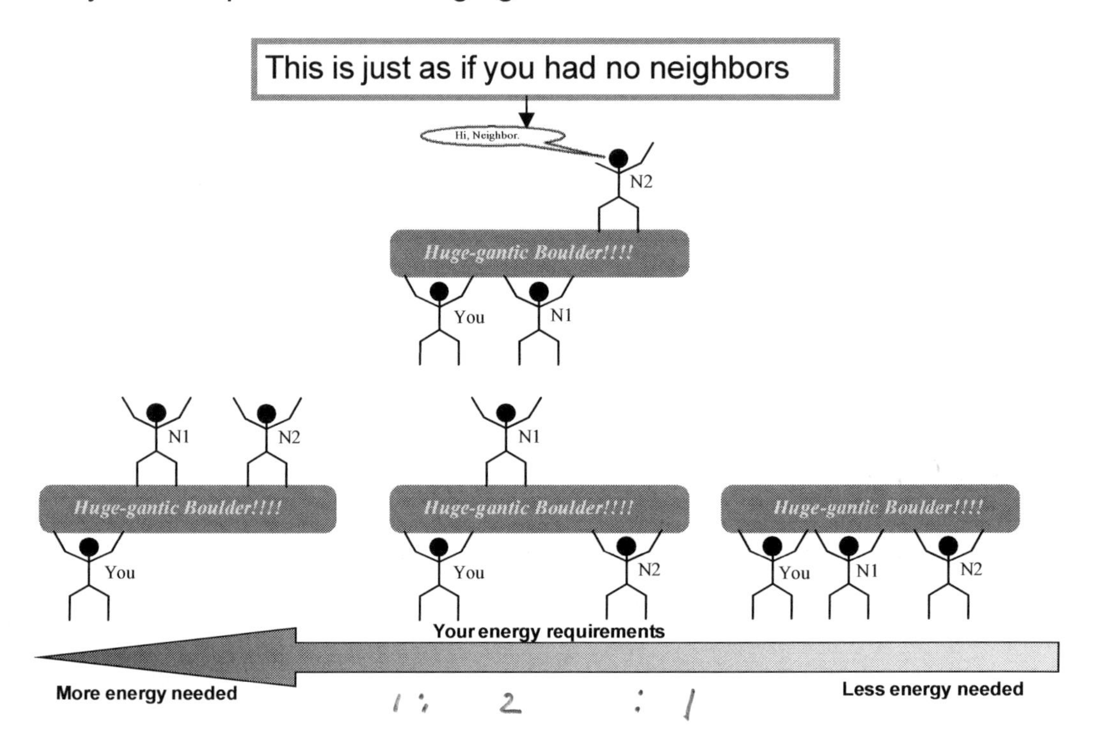

Notes:

Nuclei exhibit a statistical distribution:

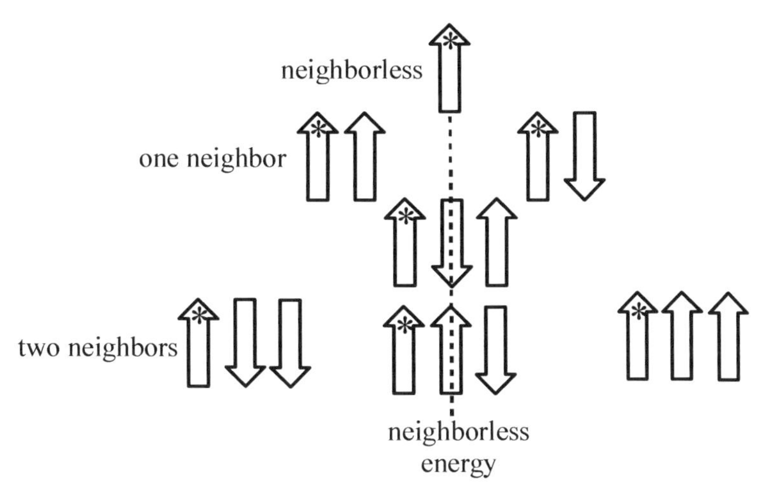

The number of peaks into which a signal is split due to neighbors is called its:

(A) multiplicity

A nucleus (or set of identical nuclei) having N neighbors on adjacent atoms has a multiplicity of:

(B) multiplicity = # of neighbors you have + 1

Notes:

Pascal's Triangle

Protons on Neighbors	Multiplicity	Relative Intensities
0	1 (singlet)	1
1	2 (doublet)	1 : 1
2	3 (triplet)	1 : 2 : 1
3	4 (Quartet)	1 : 3 : 3 : 1
4	5 (Quintet)	1 : 4 : 6 : 4 : 1
5	6 (sextet)	1 : 5 : 10 : 10 : 5 : 1

Notes:

Parts of a ^{1}H NMR Spectrum

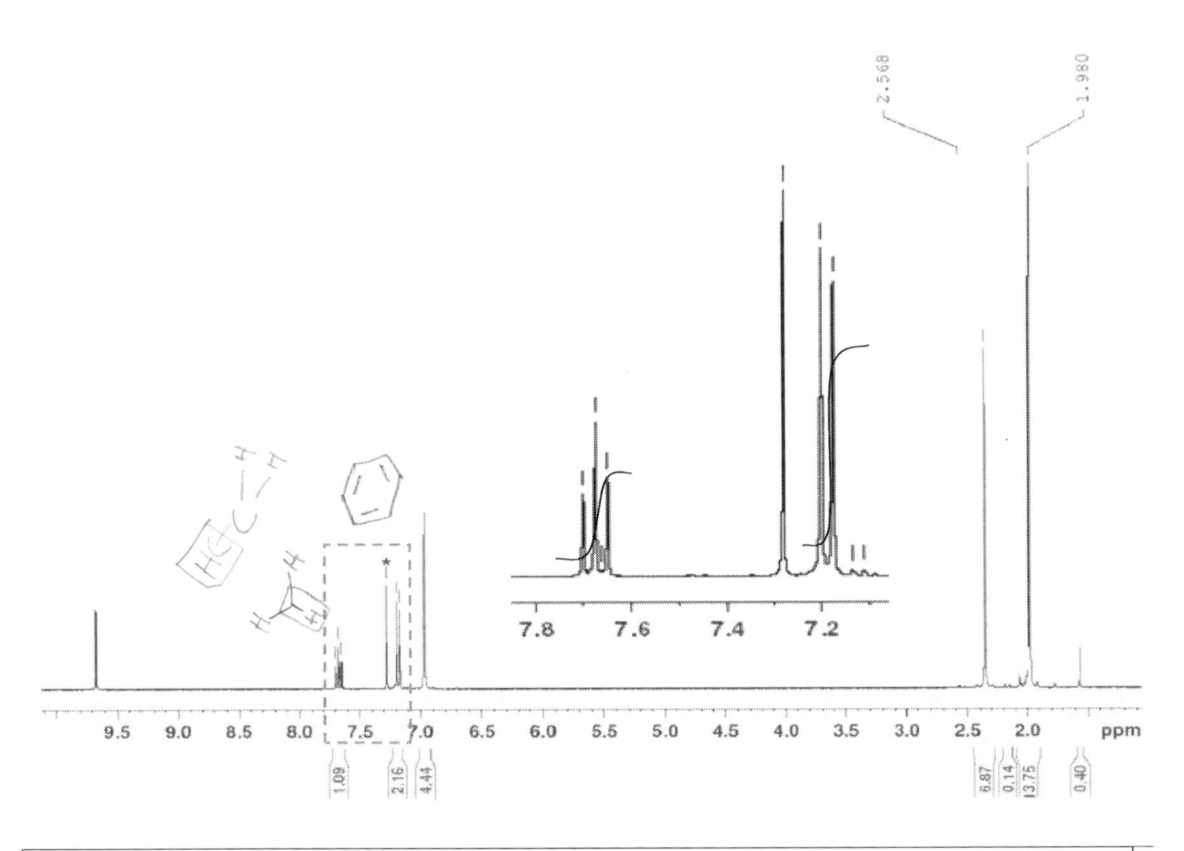

Notes:

Parts of a NMR Spectrum

Chemical Shift:

Ⓐ where peak is on x-axis

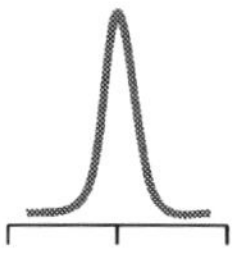

Integration:

Ⓑ area under curve or relative intensity

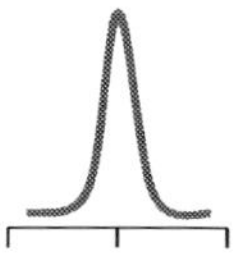

Multiplicity:

Ⓒ # of peaks in a signal

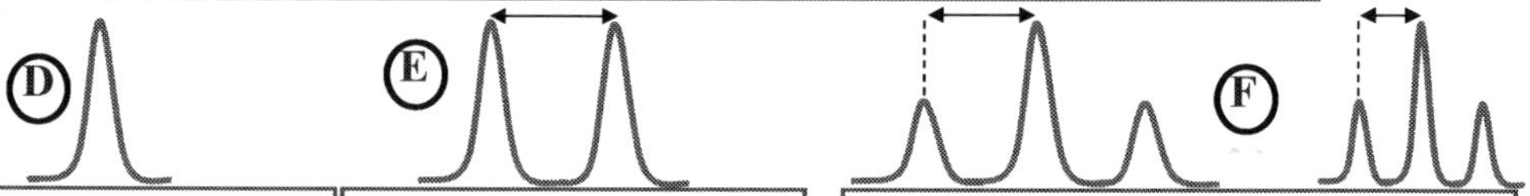

Ⓓ Ⓔ Ⓕ

Ⓓ	Ⓔ	Ⓕ
1	1 : 1	1 : 2 : 1

Coupling Constant (J):

Ⓖ The spacing b/w the peaks in a multiplet

Notes:

What do the parts of a spectrum tell us?

Chemical Shift tells us:

Ⓐ δ : tells us electronic environment about Nucleus (^1H)

Integration tells us:

Ⓑ Relative # of H's signal

Caution:
Relative Not absolute scale

Multiplicity tells us:

Ⓒ How many neighbors

Coupling Constant (J) tells us:

Ⓓ How strongly the neighboring H influences the Energy of your magnetic field

Notes:

Break it Down

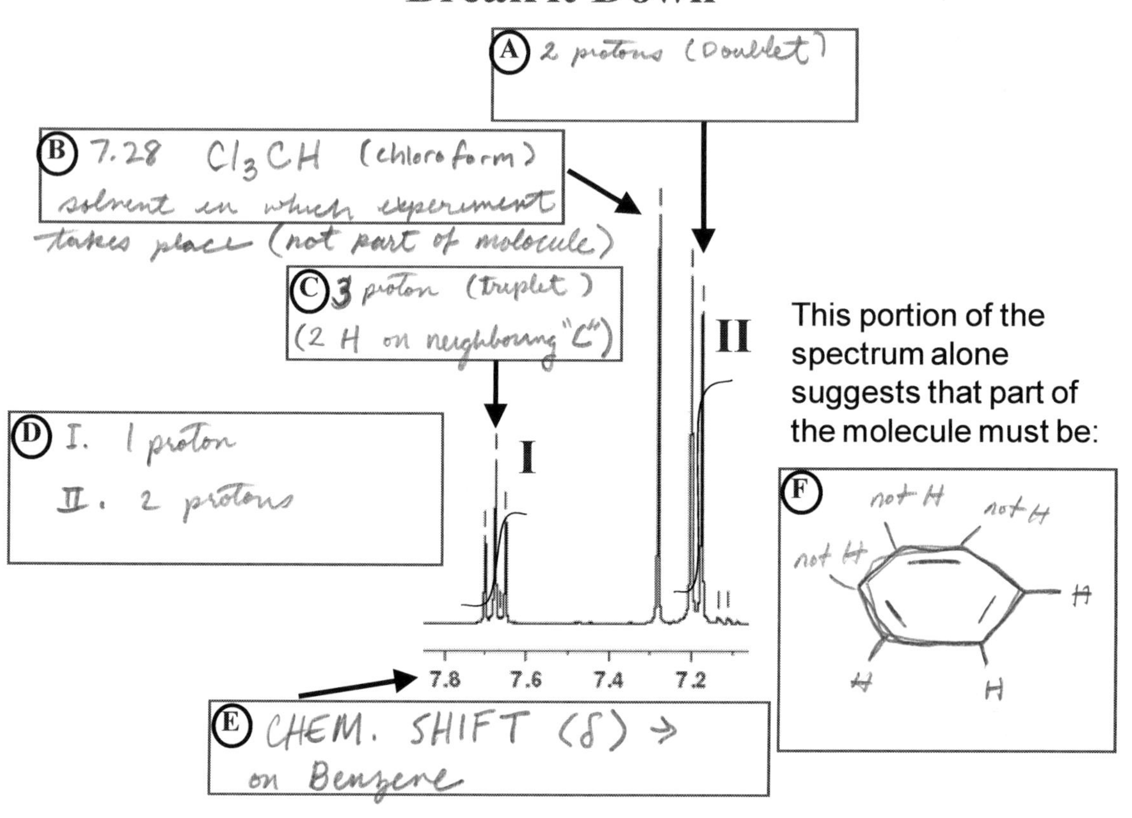

A 2 protons (Doublet)
B 7.28 Cl3CH (chloroform)
solvent in which experiment takes place (not part of molecule)
C 3 proton (triplet)
(2 H on neighboring "C")
D I. 1 proton
II. 2 protons
I
II
7.8
7.6
7.4
7.2
E CHEM. SHIFT (δ) →
on Benzene
This portion of the spectrum alone suggests that part of the molecule must be:
F
not H
not H
not H
H
H
H

Notes:

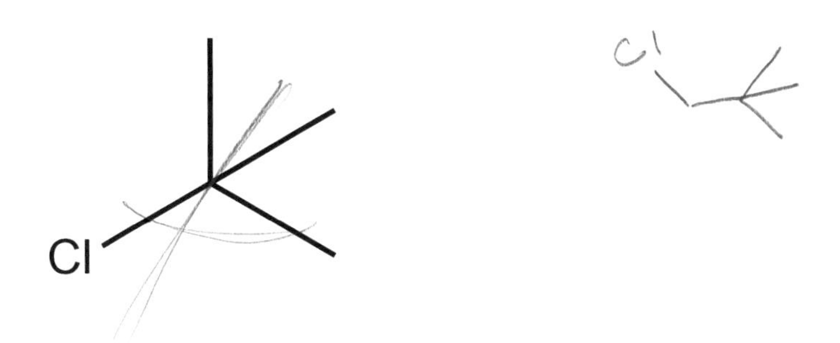

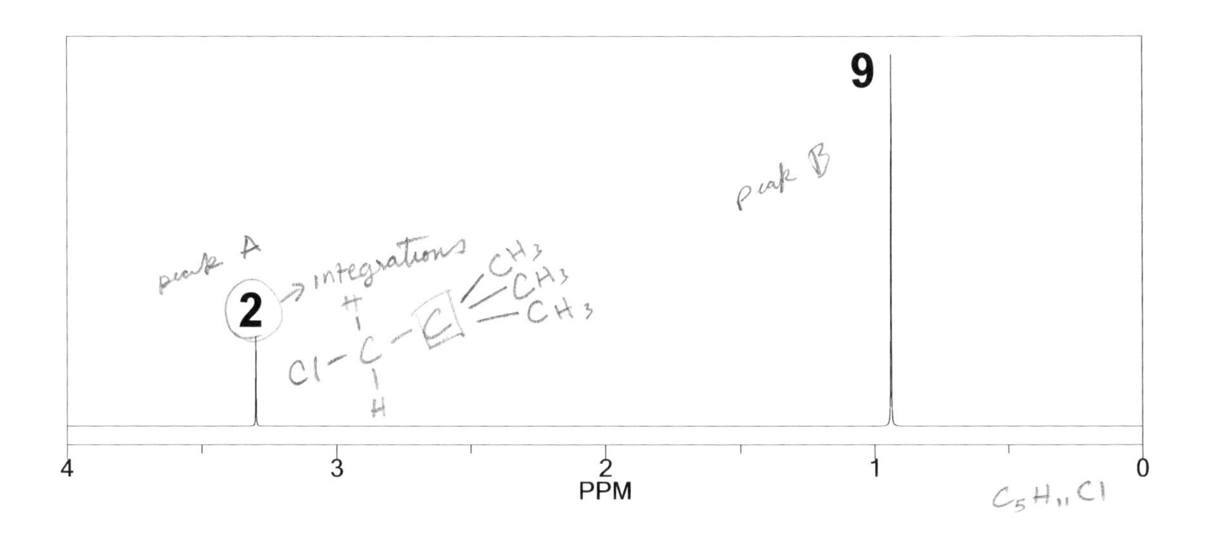

Notes:

δ: (A) "H" on same "C" as "Cl"

(B) "H" far from "Cl" (seperated by 2 or more atoms)

Int. : (A) CH_2 or CH_3 x 3

Diamagnetic Anisotropy

π electrons can move, generating local magnetic fields:

The result is:

(A)

Notes:

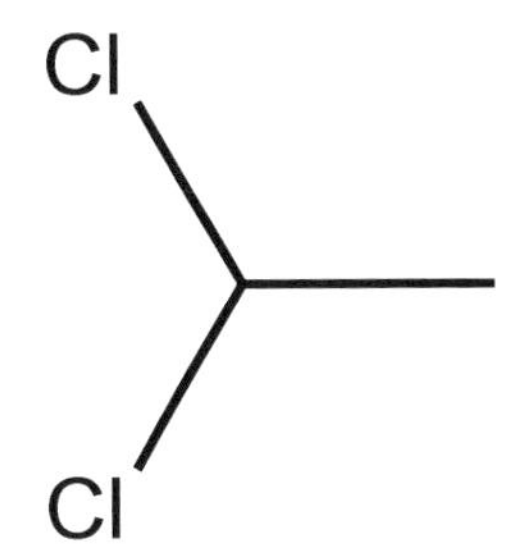

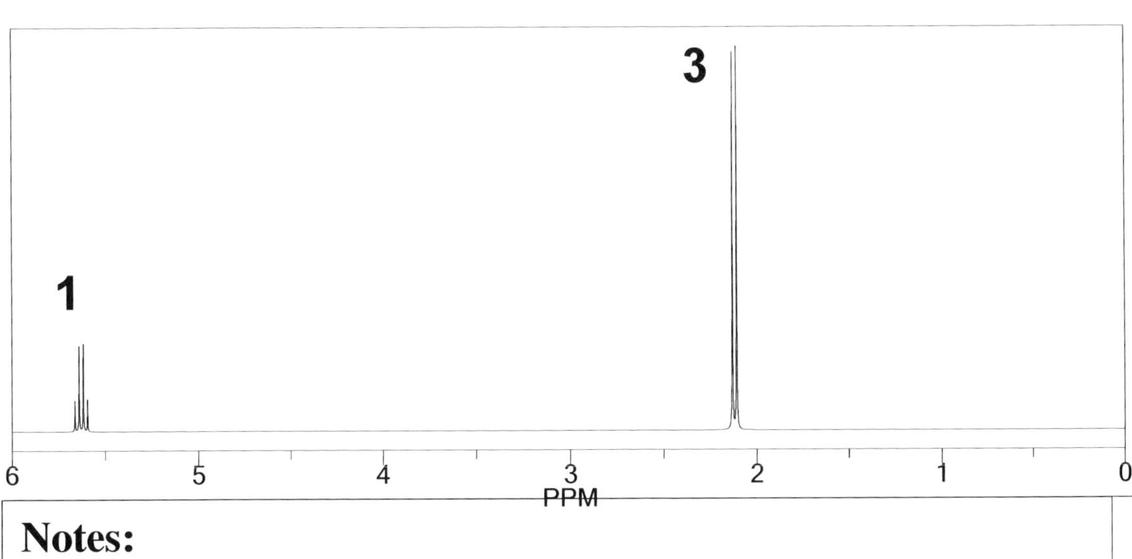

Notes:

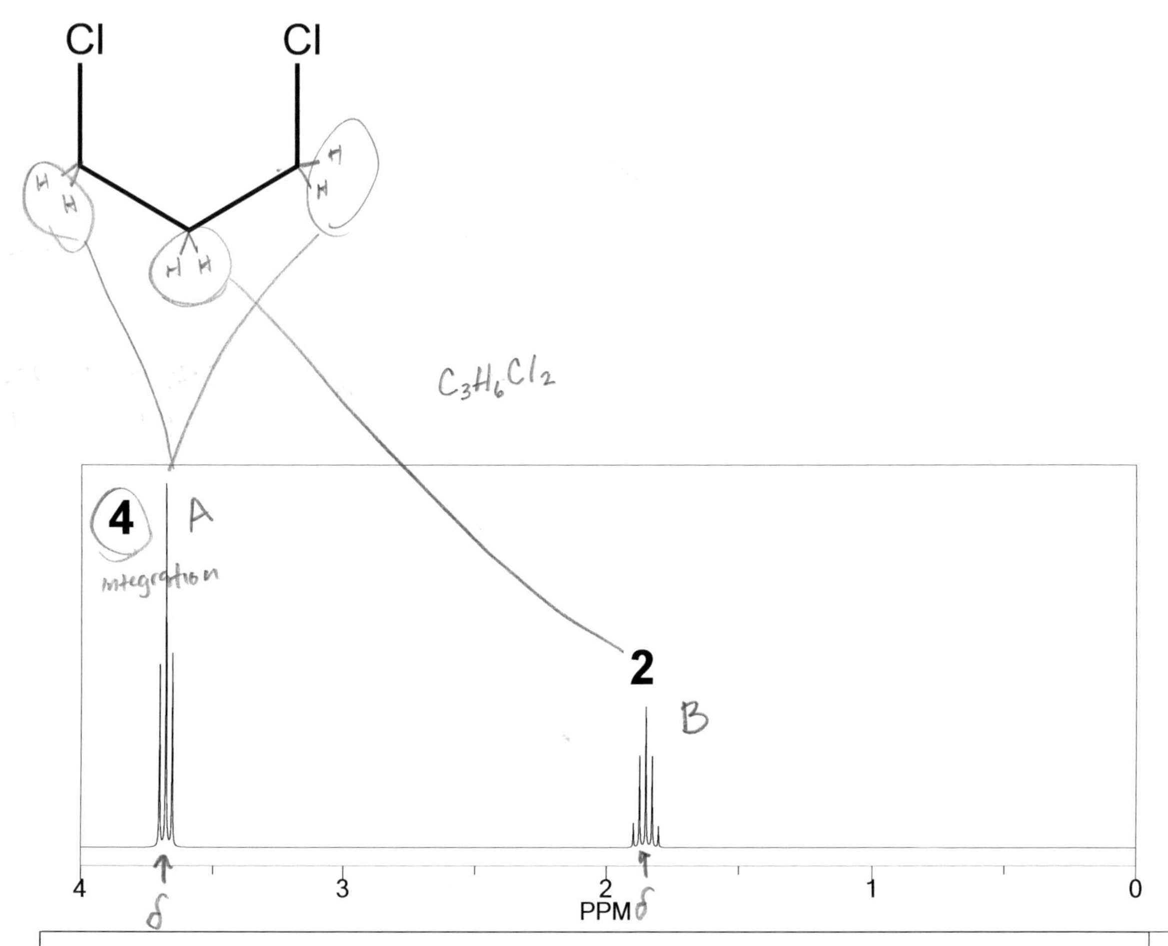

Notes:

δ: Ⓐ H on same C as Cl Ⓑ H on C next to a C w/ EN element

Int.: Ⓐ 4 so not CH_4 but 2 × CH_2 Ⓑ 2 so 1 × CH_2

Mult.: Ⓐ 2 Hydrogens on neighbor C
Ⓑ 4 " " " "

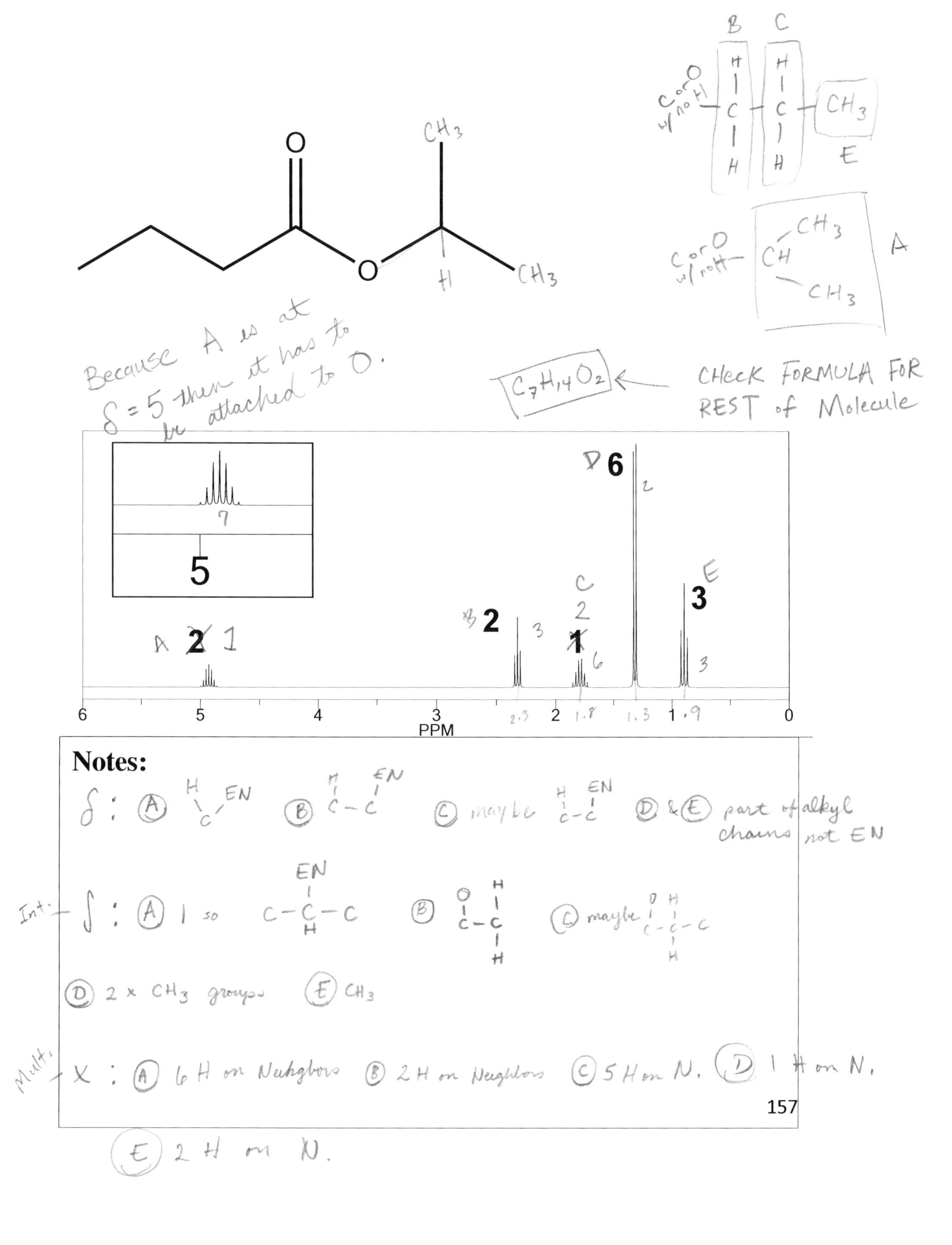
O
O
CH3
H
CH3
B
C
H
C
H
H
C
H
CH3
E
C or O w/ no H
CH3
CH
CH3
A
C or O w/ no H
Because A is at δ = 5 then it has to be attached to O.
C7H14O2
CHECK FORMULA FOR REST of Molecule
5
7
A 2 1
B 2
3
C
2
1
6
D 6
2
E
3
3
6
5
4
3
2
1
0
PPM
2.3
1.8
1.3
1.9
Notes:
δ : A H C EN B H C – C EN C maybe H C – C EN D & E part of alkyl chains not EN
Int. ∫ : A 1 so C – C – C EN H B O C – C H H C maybe O H C – C – C H
D 2 x CH3 groups E CH3
Mult. X : A 6 H on Neighbors B 2 H on Neighbors C 5 H on N. D 1 H on N.
E 2 H on N.

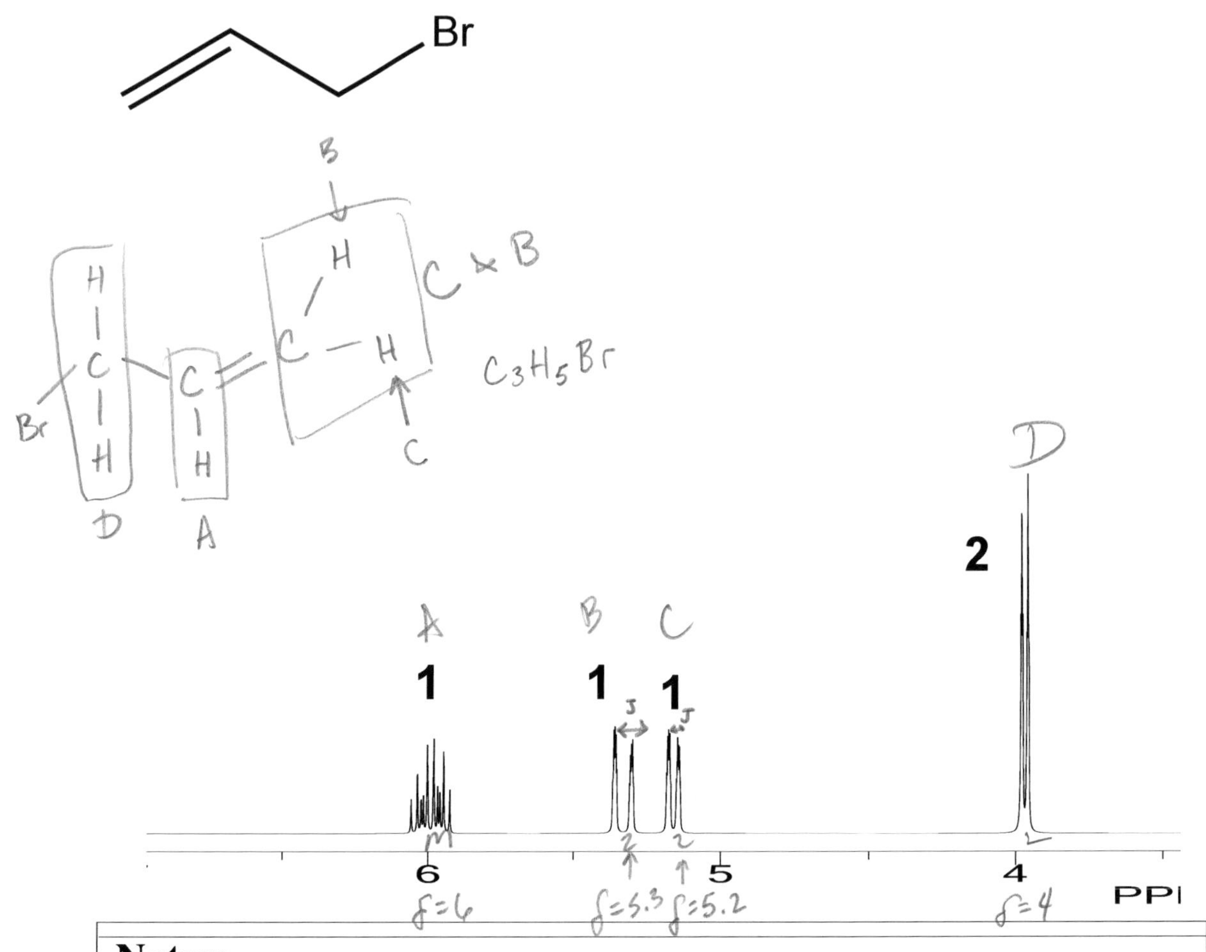

Notes:

δ: (A) next to EN element (Br) or next to or on C=C
(B) " " " "
(C) " " " "
(D) " " " "

δ: (A),(B),(C) = CH & (D) CH_2

x: (A) ? (B) 1 H on N.
(C) " " "
(D) " " "

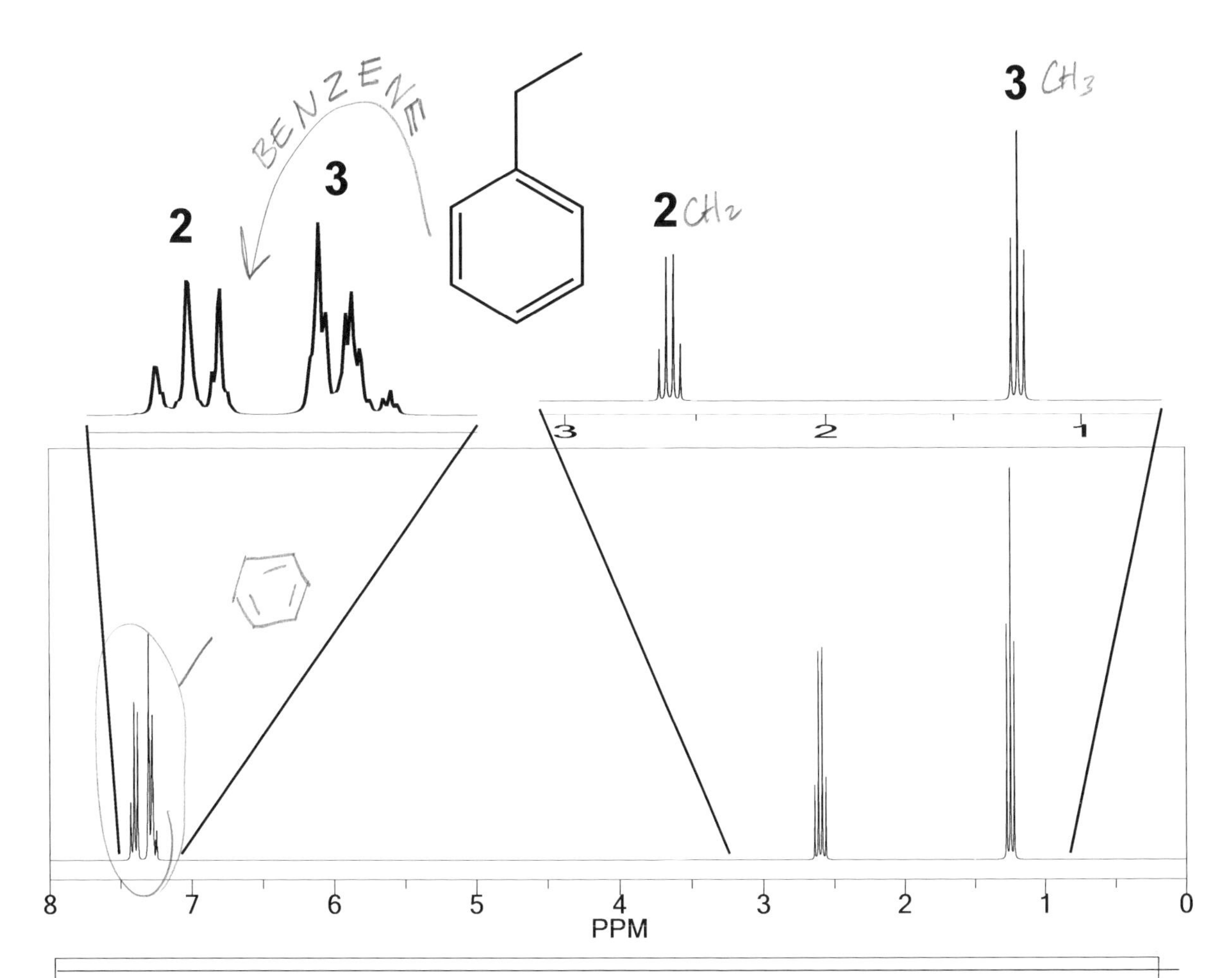

Notes:

* peaks b/w 7-8 on X-axis think BENZENE

ALWAYS OVERLAP

DON'T

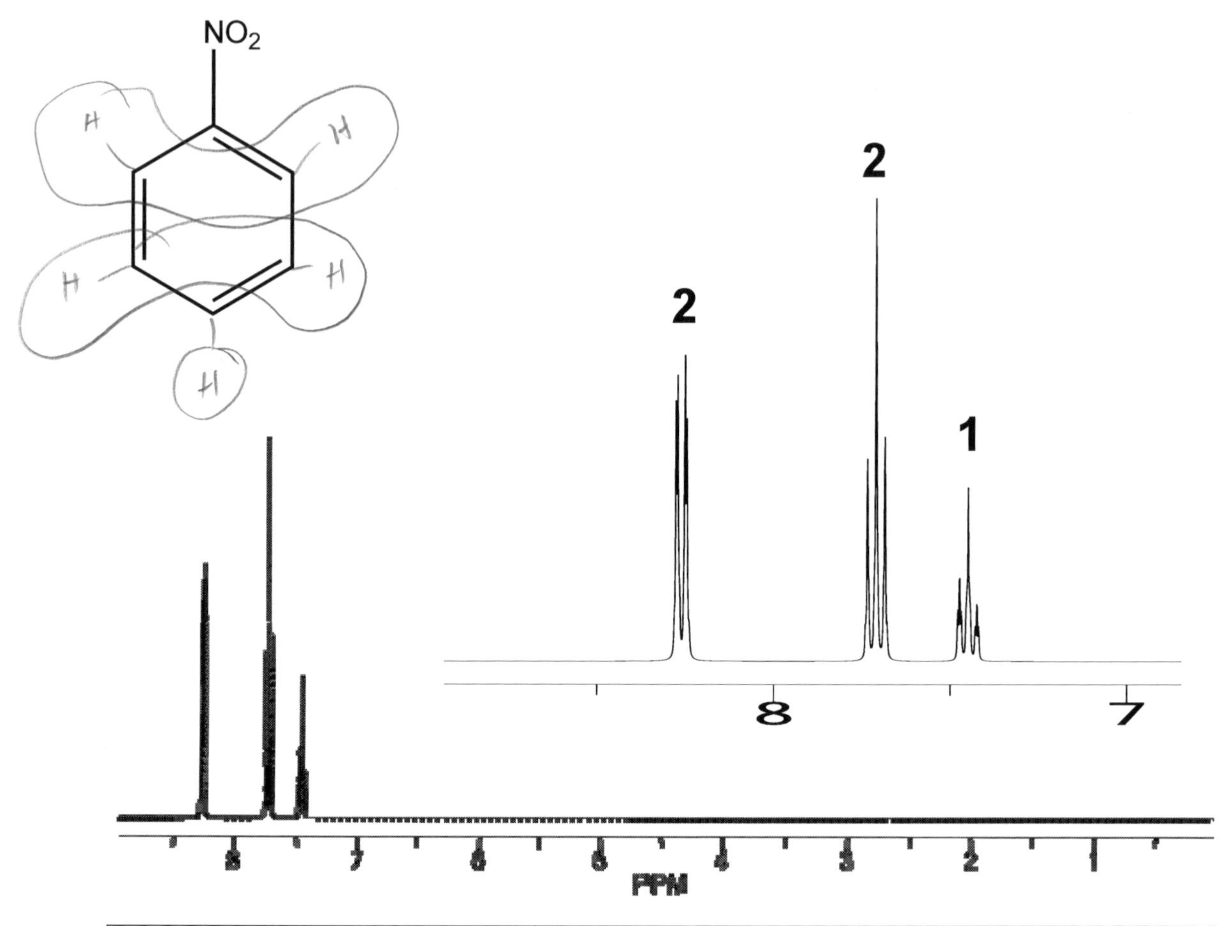

Notes:

Splitting by inequivalent neighbor sets I: doublet of doublets

Notes:

Splitting by inequivalent neighbor sets II: quartet of triplets

Notes:

Effect of H-Bonding

Recall that in IR spectra, one observes broad signals and concentration / water content for bonds involving H-bonding protons. Effects of this nature are also important to consider for NMR spectra:

(A)

(B)

Notes:

^{13}C NMR Spectroscopy

Protons are not the only nuclei that are **NMR active** (capable of exhibiting nuclear magnetic resonance in a magnetic field). Organic molecules all contain carbon atoms, so probing carbon by NMR is very useful. Carbon-12 nuclei are **not** NMR active, but the less-abundant isotope ^{13}C is.

Thus, ^{13}C NMR is a useful technique for determining how many different kinds of carbons a molecule has. Because only the ^{13}C isotope is NMR active, we DO NOT see splitting of signals (the H-induced splitting is removed by **decoupling** done by the instrument settings).

TMS is still the reference (chemical shift = 0 ppm).

Notes:

Carbons (Shown)	Chemical Shift
$Si(CH_3)_4$	0
$—CH_3$	7-35
$—CH_2—$	15-50
 $—CH(—)—$	20-60
 $—C(—)(—)—$	30-40
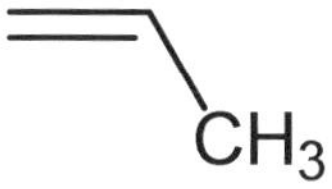 $=CH—CH_3$ (alkene)	100-150
 aromatic CH	100-175

Carbons (Shown)	Chemical Shift
$—C—X$, X = I	0-40
X = Br	20-60
X = Cl	30-80
X = O	50-80
$—C(=O)X$, X = R	200-220
X = H	190-200
X = OH	175-185
X = OR	165-175
X = NR_2	160-175
 $—C\equiv C—$	65-100

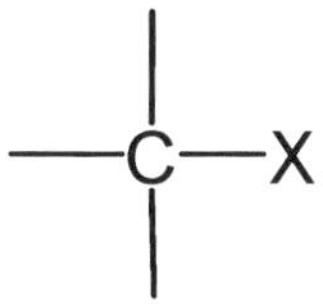

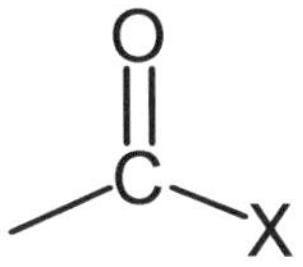

Notes:

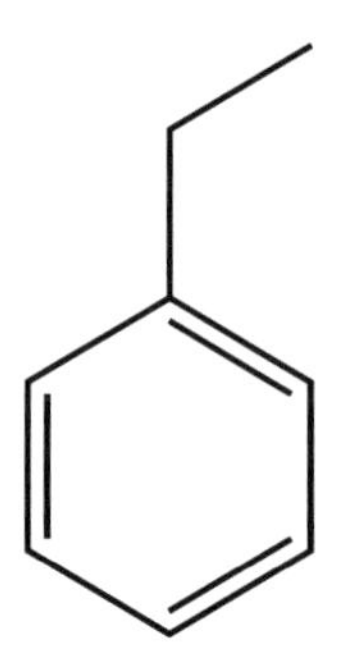

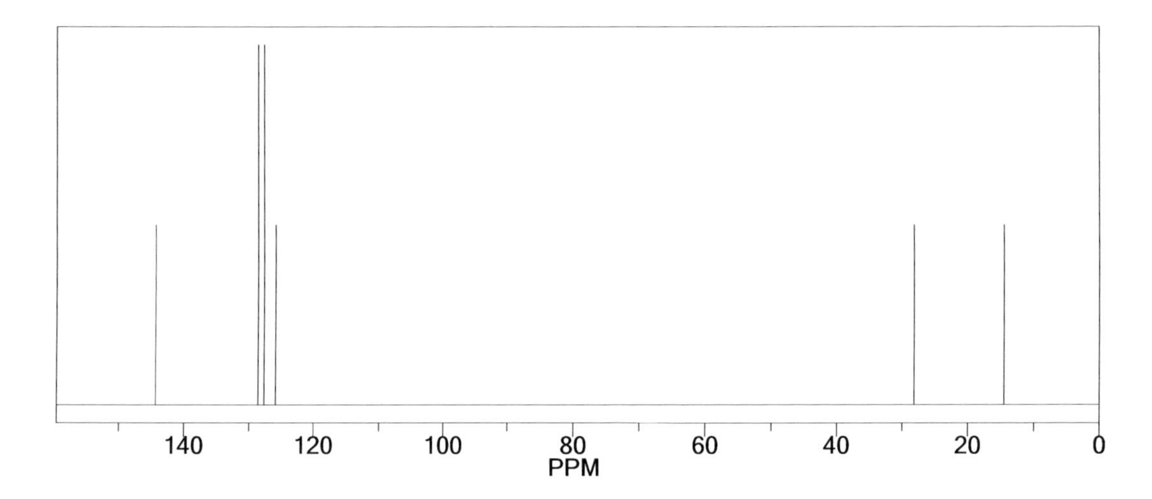

Notes:

Lecture Set 4:
Aromaticity and Reactions of Benzene

Suggested Reading:

Suggested Problems:

Recap: π-Conjugation

This lecture set will delineate special properties of molecules endowed with **alternating single and double bonds, i.e.**:

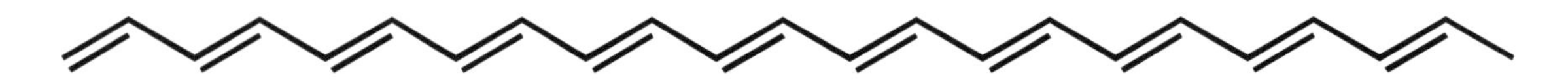

Ultimately, the special properties and reactivities of such systems derive from

Ⓐ Delocalization of e^- in π bonds (p orbitals)
(Resonance Stabilization)

Notes:

- All atoms in conjugated systems or Rings → sp^2

Benzene is Special

Description: C_6H_6; six sp^2 hybridized carbons in a planar hexagonal cycle

Observations: C-C-C angles: 120º

C-C bonds: 1.40 Å

Structure:

(A)

Notes:

- Resonance of Benzene contributes to extreme stability

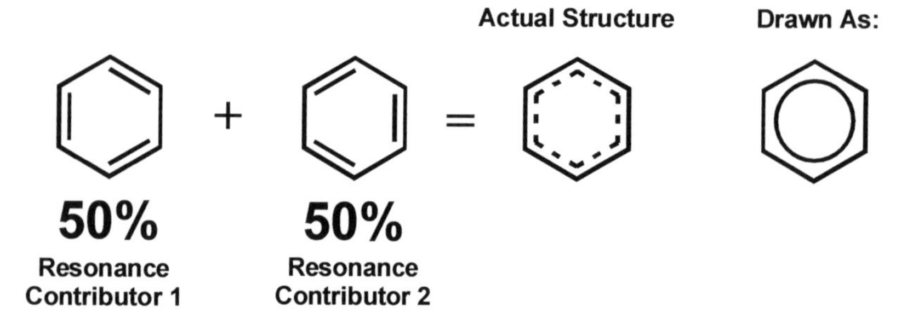

Pi-electron system:

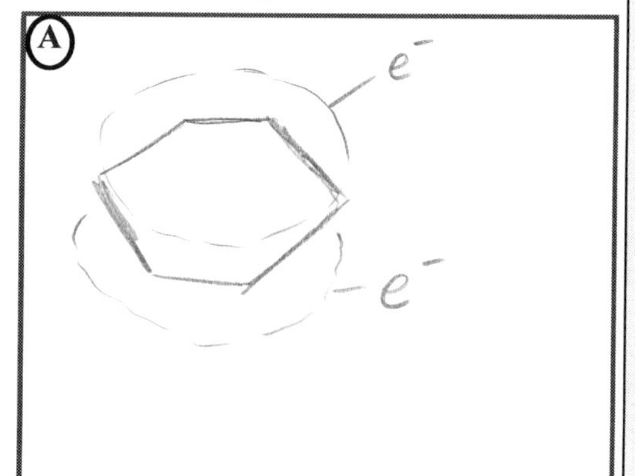

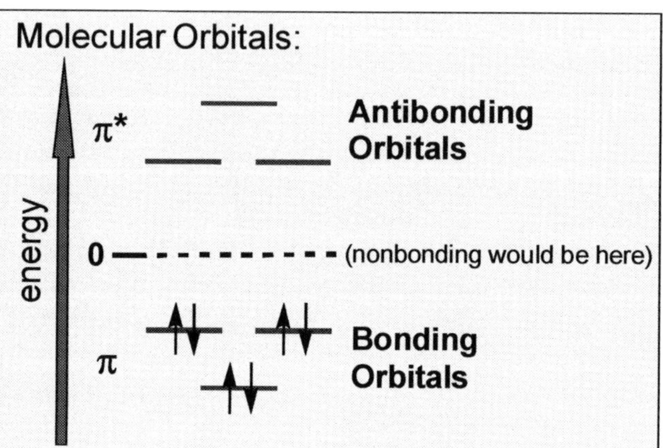

Notes:

Stabilization of Benzene

Reaction	Calc. Heat Hydrog (kJ/mol)	Observed	Delocalization Energy
cyclohexane $\xrightarrow{H_2}$ cyclohexane		119.7	0
benzene $\xrightarrow{3H_2}$ cyclohexane	3(119.7) = –359.1	208.4	**151**
cis-1,3,5-hexatriene $\xrightarrow{3H_2}$ hexane	3(119.7) = –359.1	336.8	22.3

Aromatic stability

π conjugation

Benzene is about 150 kJ/mol more stable than would be expected; this special type of stabilization, is known as (A) Aromaticity

Compounds having this stability are called **aromatic** compounds, and are said to possess **'aromaticity'**.

Notes:

Criteria for Aromaticity

In order to exhibit aromaticity, a compound must:

1. Planar AND CYCLIC
2. Uninterrupted π system (no sp^3 atoms)

3. Conform to Hückel's Rule:

$4n+2$ π e^-

n = 0: 2 electrons
n = 1: 6 electrons (i.e., benzene)
n = 2: 10 electrons

WHAT IS 'n'??? is any integer

a planar compound with an uninterrupted pi system having **4*n* electrons is said to be "antiaromatic"**

Notes:

Aromaticity:

$4n+2$ π e^-

Cyclic

Planar

unitterrupted π system (no sp^3)

Aromatic?

Examine some compounds to assess whether they are aromatic:

Cyclopropene

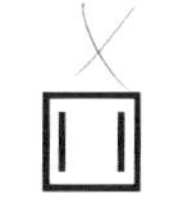

Cyclobutadiene

Benzene

Cyclopenta-1,3-diene

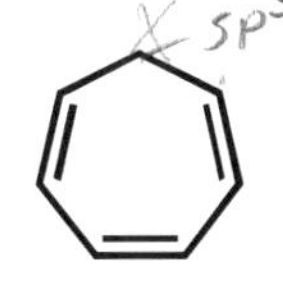

Cyclohepta-1,3-diene

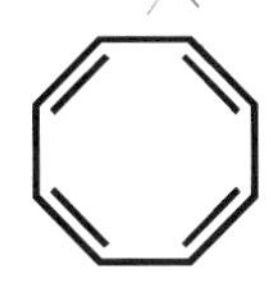

Cyclooctatetraene

4n + 2 ?

Uninterrupted ?

Notes:

Benzene ONLY

Aromatic?

Consider these charged species. Assuming all are planar and have an uninterrupted pi cloud (sp^2-hybridized atoms), check for Hückel's rule condition:

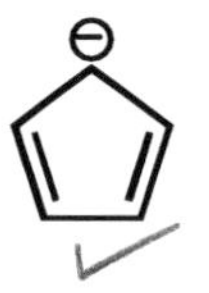
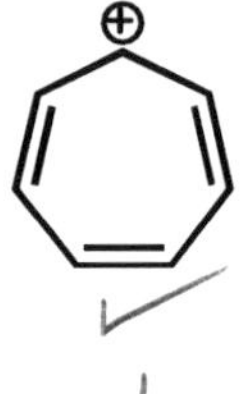
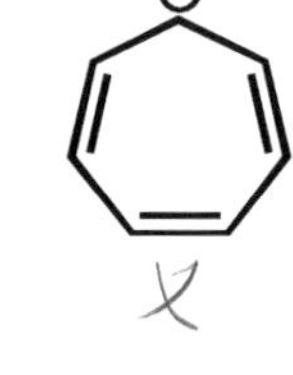

4n+2?

Tropilium Cation

A note on lone pairs

IF:

THEN:

Notes:

(+) → sp^2 (carbocation)

(−) → Normally sp^3, however, ...

IF: Putting lone pair in "p" orbital makes system AROMATIC THEN put lone pair in "p" orbital

Polycyclic Systems

What if you have a polycyclic system?

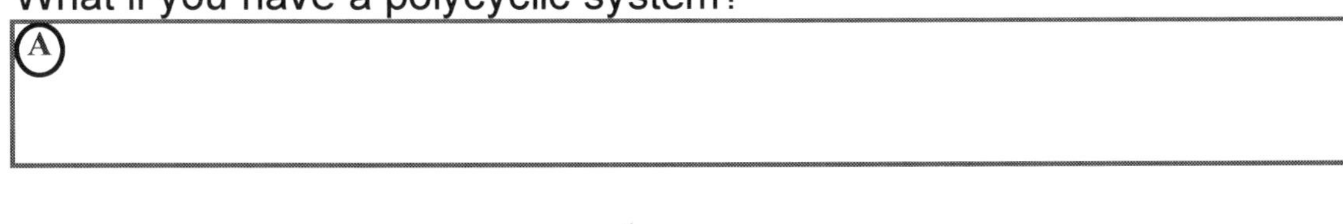

10e⁻ ✓

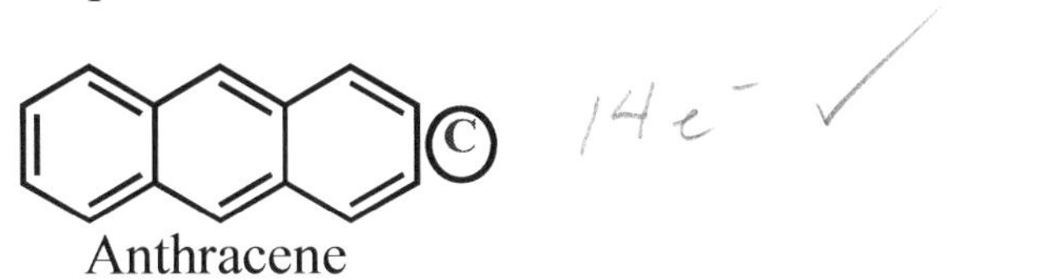

14e⁻ ✓ 4n+2 ?

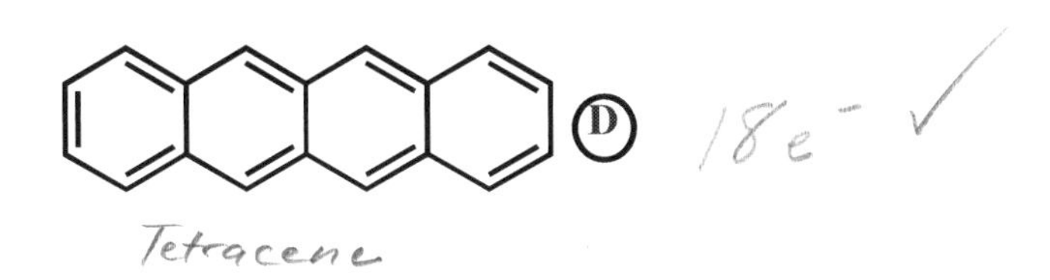

18e⁻ ✓

Tetracene

Notes:

Aromatic Heterocycles (having non-carbon atoms)

What if a ring has heteroatoms (any non-carbon atom) in it?

(A)

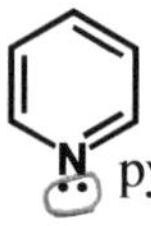

pyridine ✓ Nitrogen ⇒ sp^2

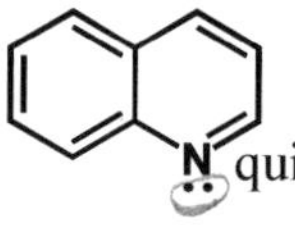

quinoline ✓

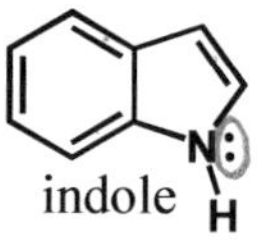

indole ✓ Nitrogen ⇒ sp^3 ⇒ lone pair must go in "p" orbital

* count lone pair *

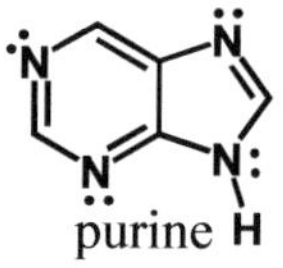

purine

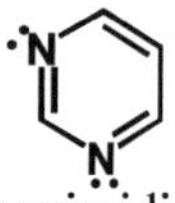

pyrimidine ✓

Notes:

Aromatic Heterocycles: Episode 2

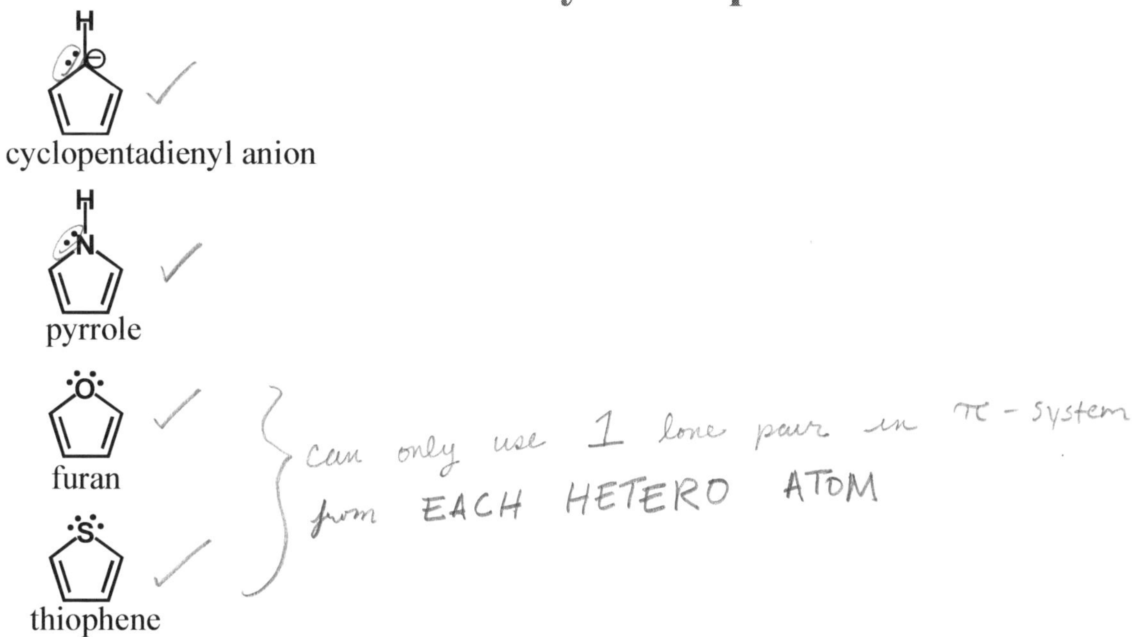

<u>Molecules tend towards the most stable (lowest energy) state***</u>

Notes:

Aromaticity: Effects on Basicity

The properties of molecules can be predicted by thinking about the behavior of the lone pairs as related to aromaticity:

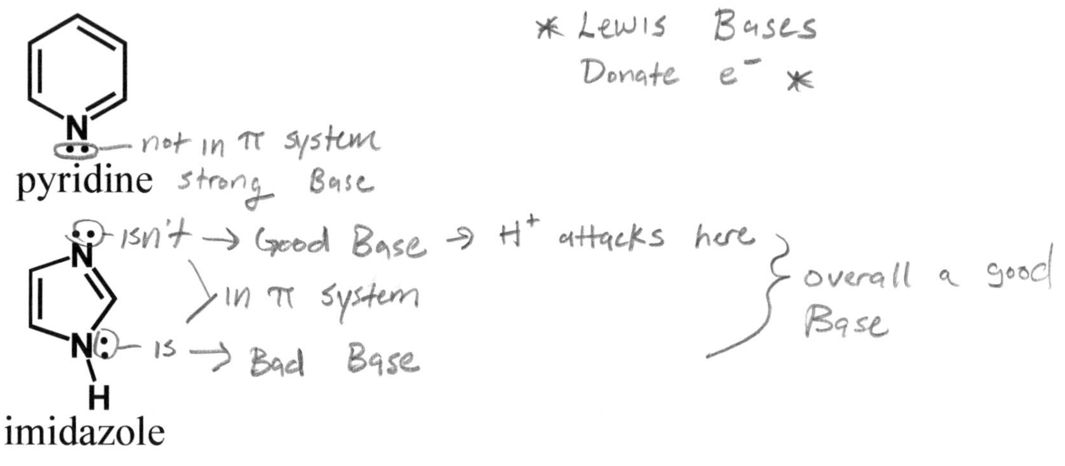

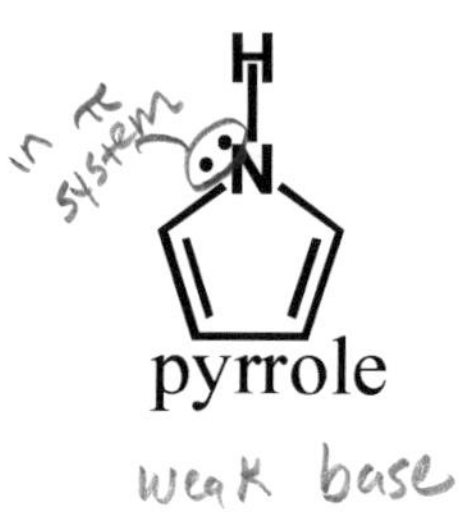

Notes:

VERY IMPORTANT

lone pair not in π system = GOOD BASE

Antiaromaticity

What if a ring has the 'wrong number' of electrons in its pi system?

A

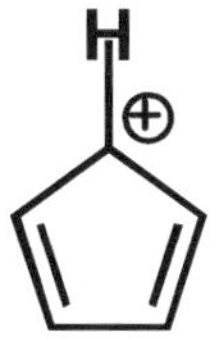

cyclopentadienyl cation

SKIP

cyclobutadiene

Notes:

Nomenclature – Monosubstituted Benzene

The straightforward:

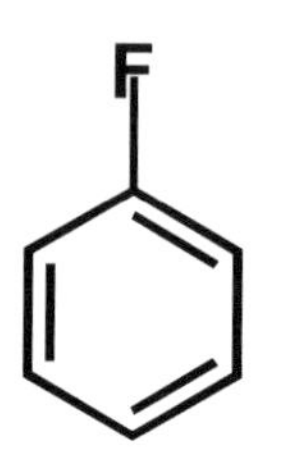

Flouro benzene

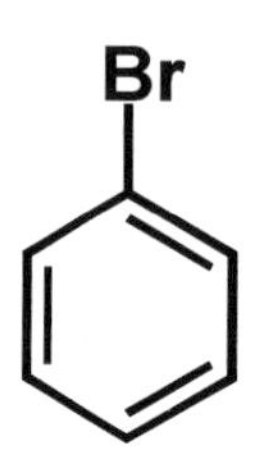

Bromobenzene

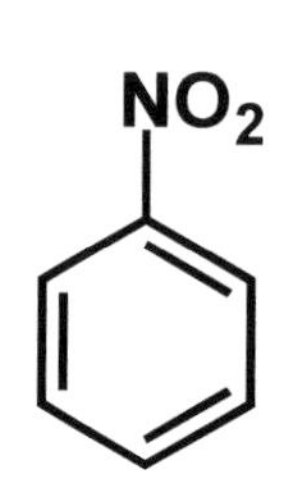

Nitrobenzene

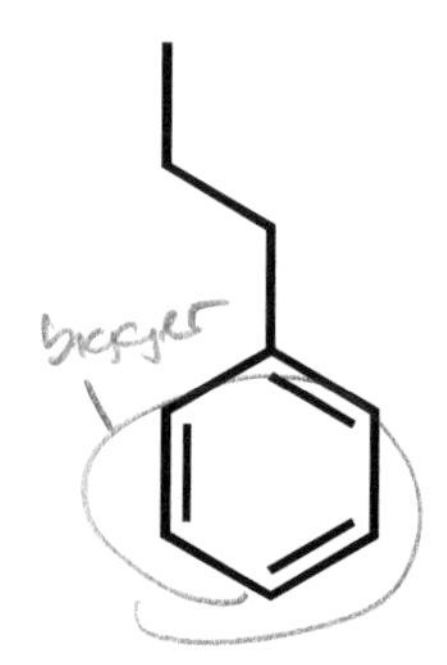

Propylbenzene

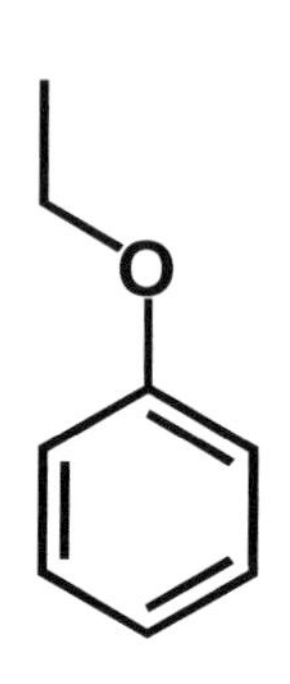

ethoxybenzene

The not-so-bad:

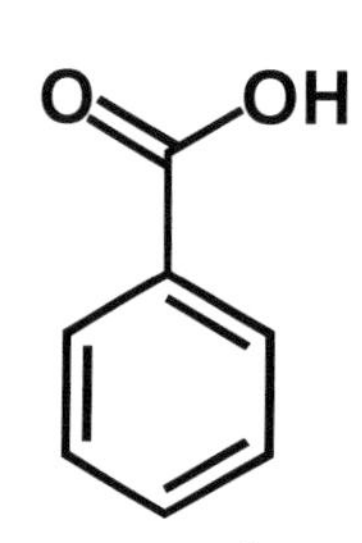

Benzoic Acid

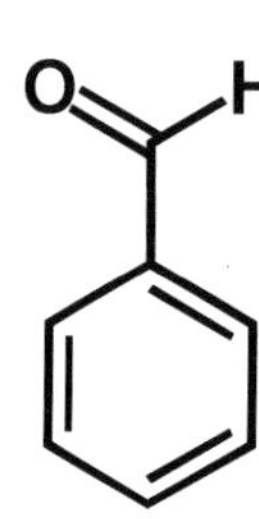

Benzeldehyde

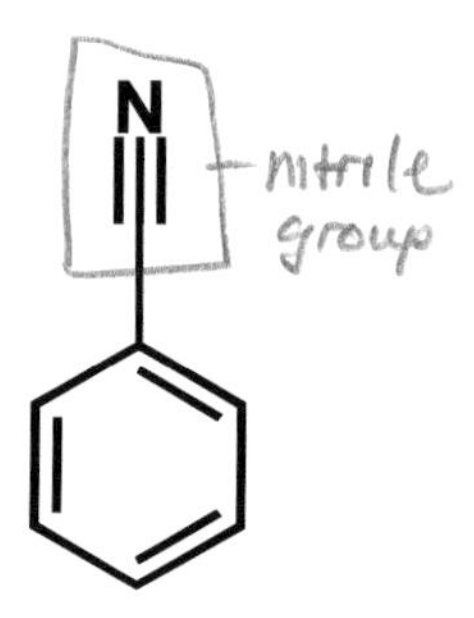

Benzonitrile

Notes:

Names Rooted in the Storied History of Chemistry

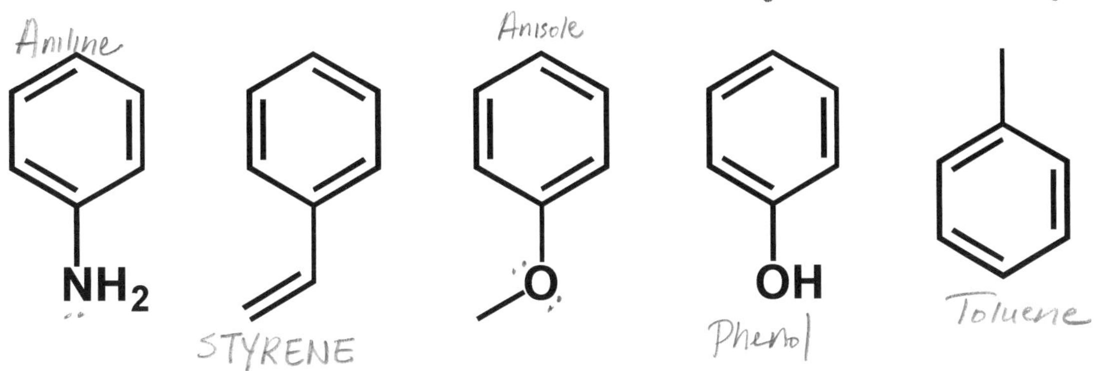

Notes:

Review: Reaction of Unspecial Alkenes

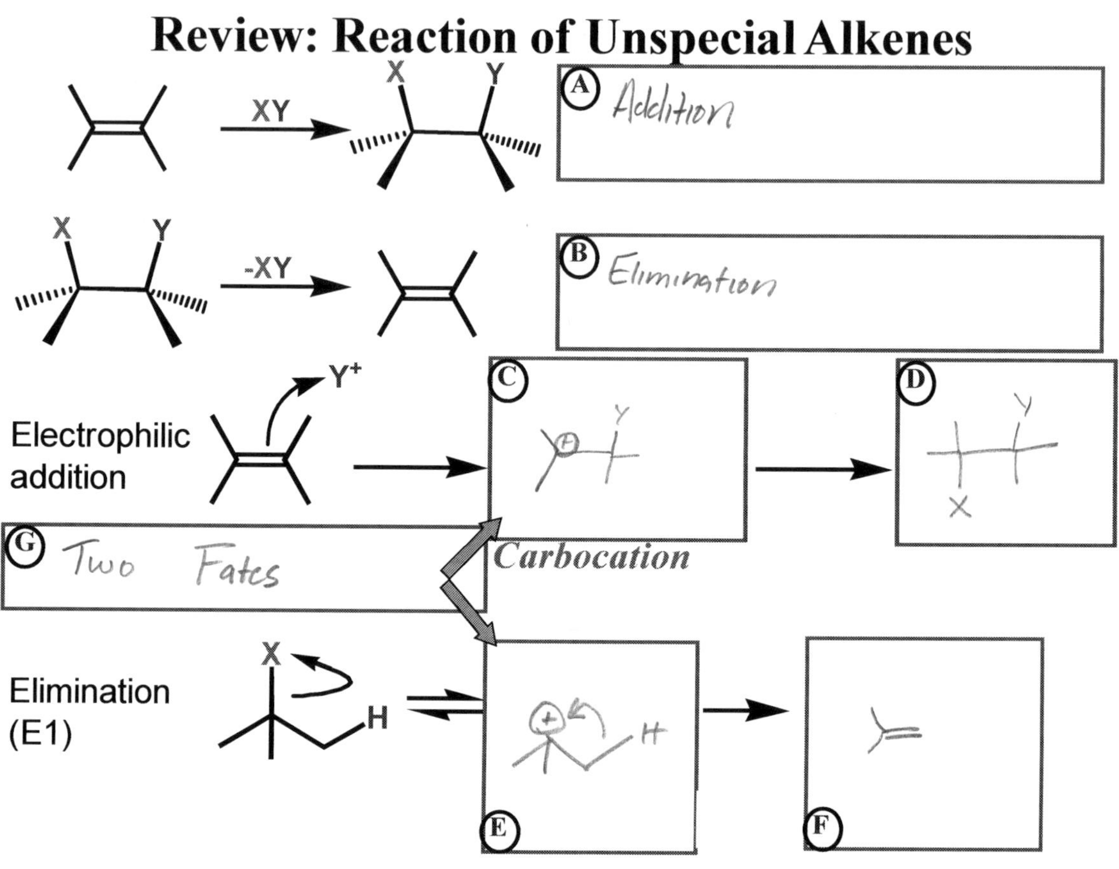

Notes:

Reaction of Benzene

Because aromatic compounds (arenes) like benzene are stabilized by aromaticity, their double bonds tend to be less reactive than alkene double bonds. However, with more forceful conditions or more reactive reagents, many reactions of aromatic compounds can occur. The example we will look at is

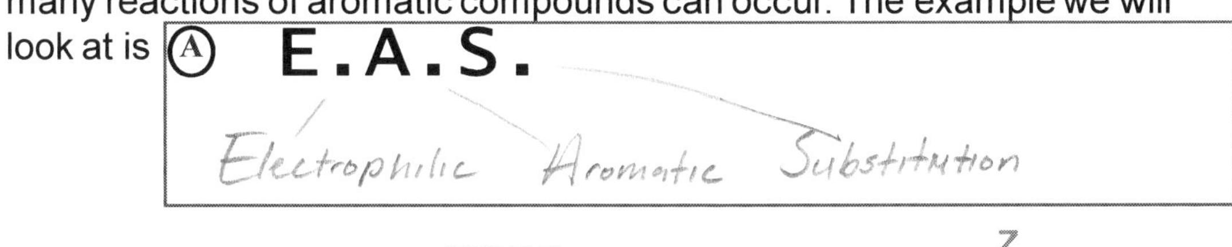

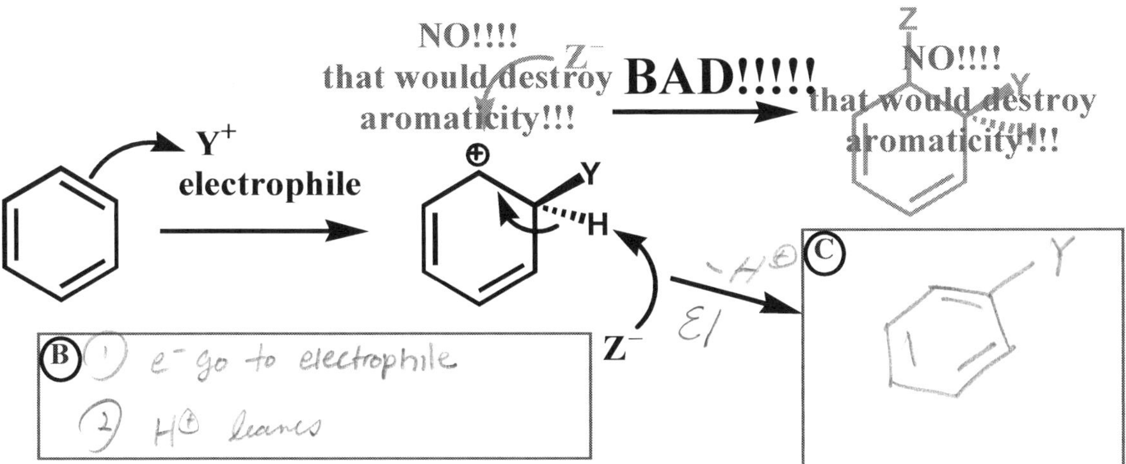

Notes:

Good News About EAS*

You will learn many electrophilic aromatic substitution reactions;

They all use the same mechanism!

1. Give electrons to electrophile (makes a carbocation)

2. Eliminate H^+ (yields the aromatic compound with the electrophile on it)

Here we see iodination of benzene:

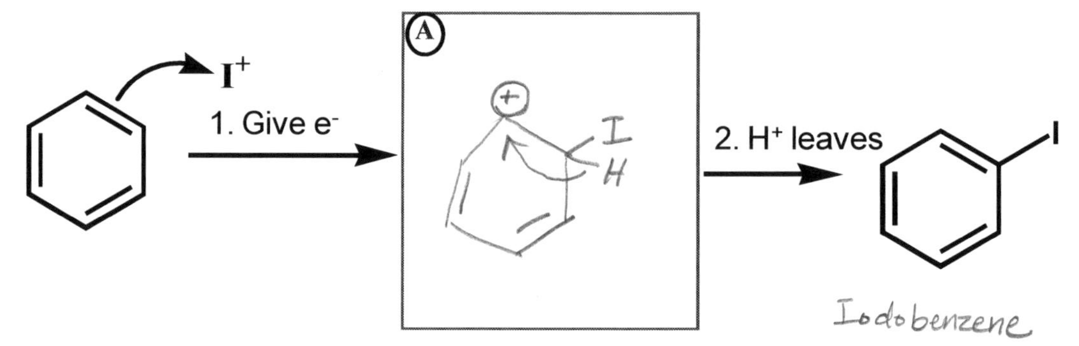

The following few slides show specific examples. The only really unique thing is that a different **ELECTROPHILE** is involved in each case, and the reactants you use lead to formation of these electrophiles in different ways...

*EAS = Electrophilic aromatic substitution

Notes:

Halogenation I: Chlorination and Bromination

I. The reaction

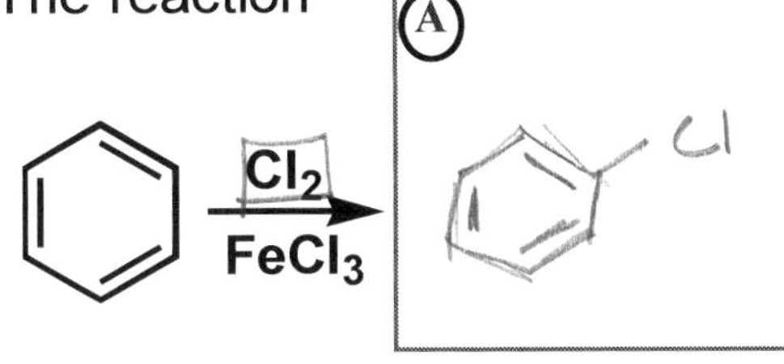

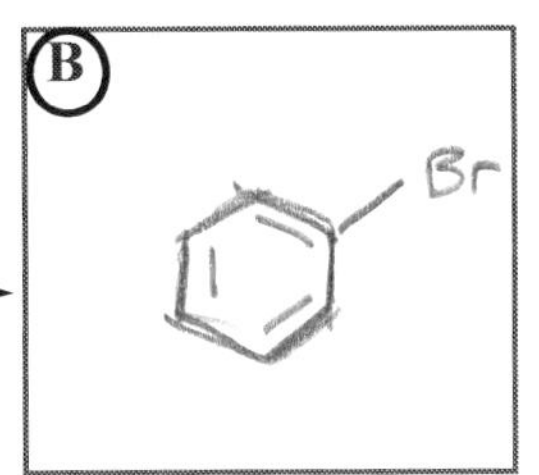

II. How is the electrophile generated?

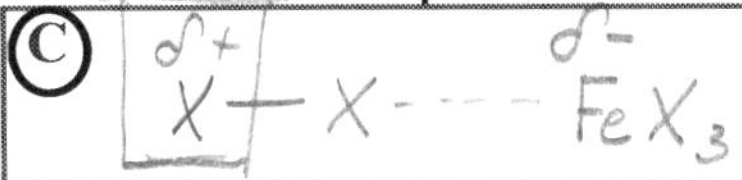

NOTE: the electrophile is not a free cation:

C δ+ X—X---- δ- FeX_3

III. The Mechanism:

+ X—X--- FeX_3 → ⊕ X H + FeX_4^- → X

Notes:

Halogenation II: Iodination

I. The reaction

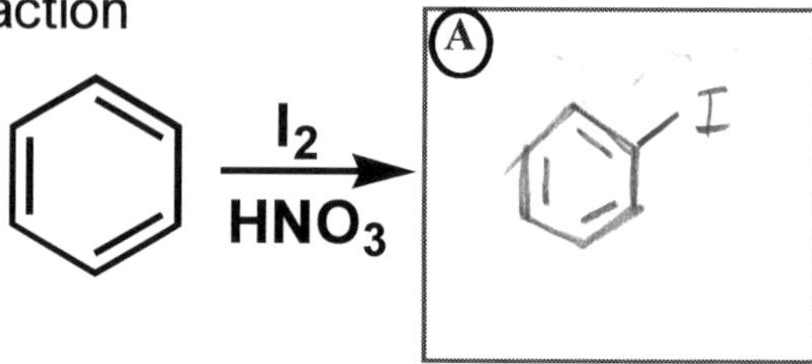

II. How is the electrophile generated?

$$I_2 + HNO_3 + 2H^+ \longrightarrow 2I^+ + HNO_2 + H_2O$$

B I_2 (ox. state = 0) [ox] → $I^{\oplus}$

III. The Mechanism:

Notes:

Nitration

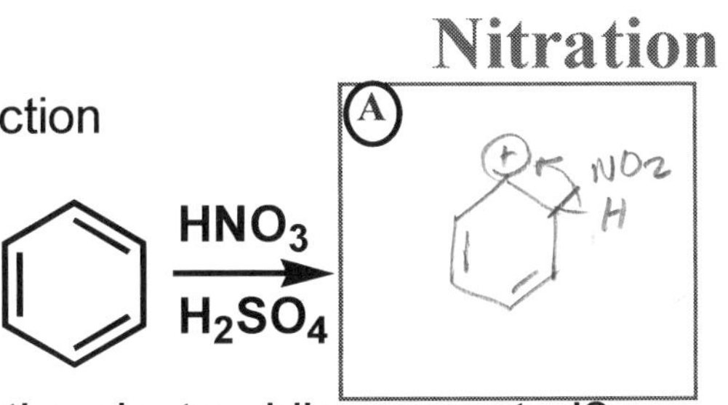

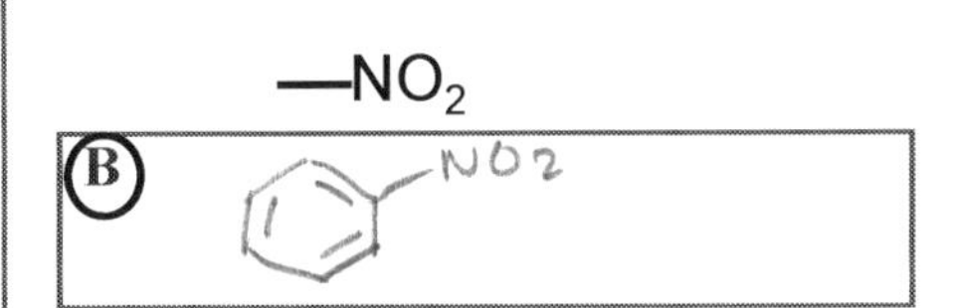

II. How is the electrophile generated?

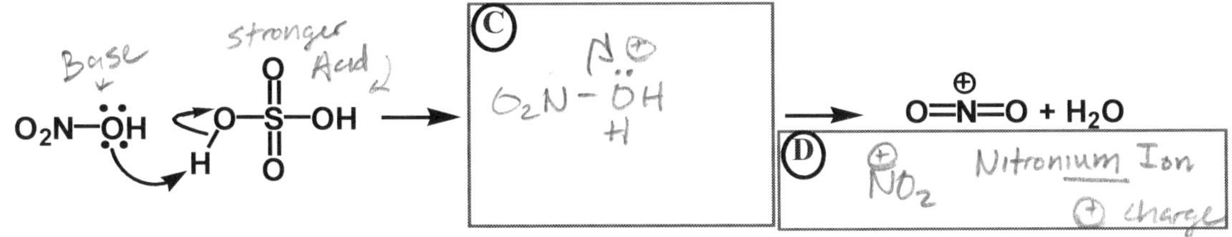

III. The Mechanism:

Notes:

Sulfonation

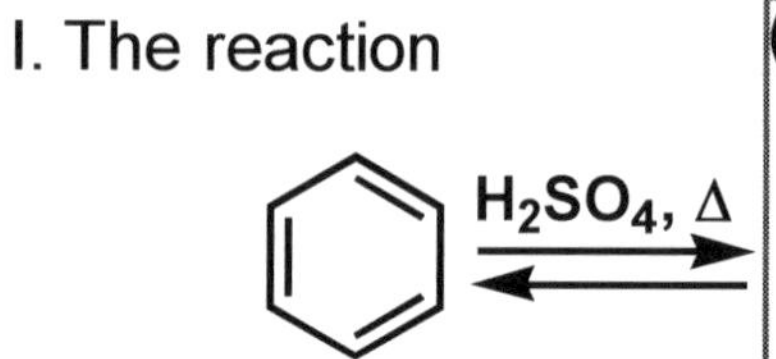

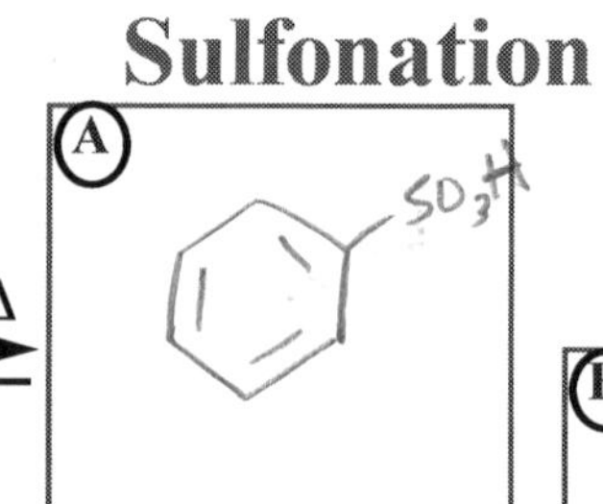

—SO_3H

(B) Sulfonic Acid

II. How is the electrophile generated?

Amphoteric

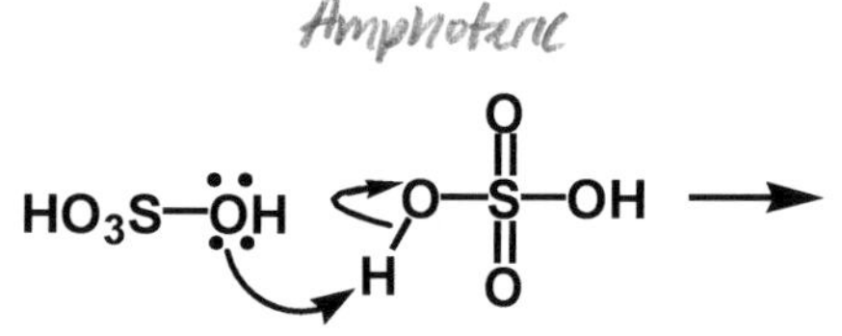

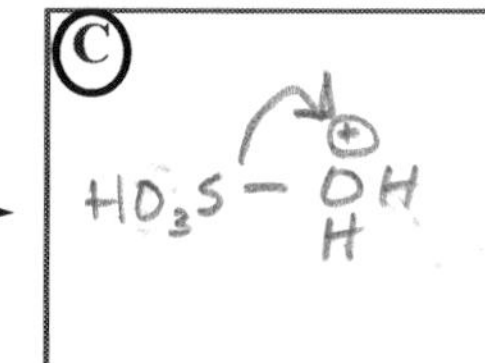

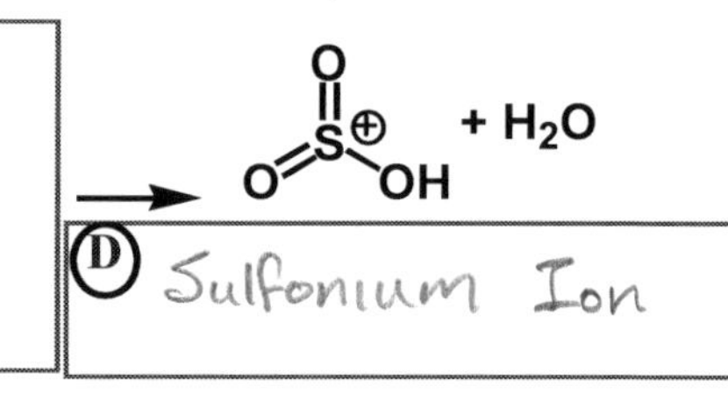

III. The Mechanism:

Notes:

Friedel-Crafts Acylation

I. The reaction

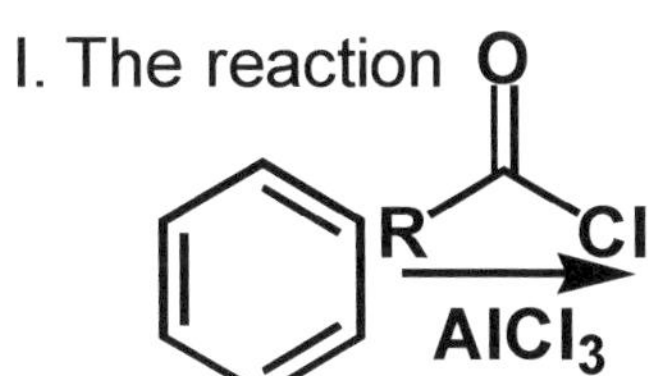

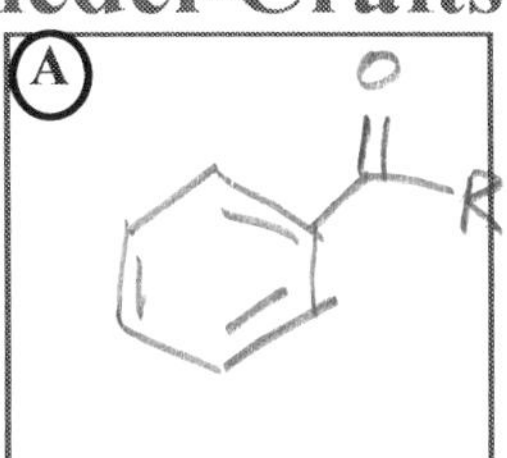

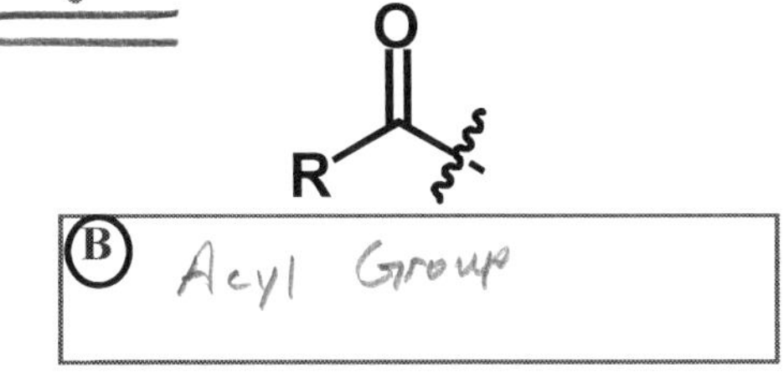

II. How is the electrophile generated?

R–C(=O)–Cl + Cl–AlCl₂ → C: R–C⊕=O + AlCl₄⁻

R—C⊕═O

D: Acylium ion

III. The Mechanism:

Notes:

Friedel-Crafts Alkylation

I. The reaction

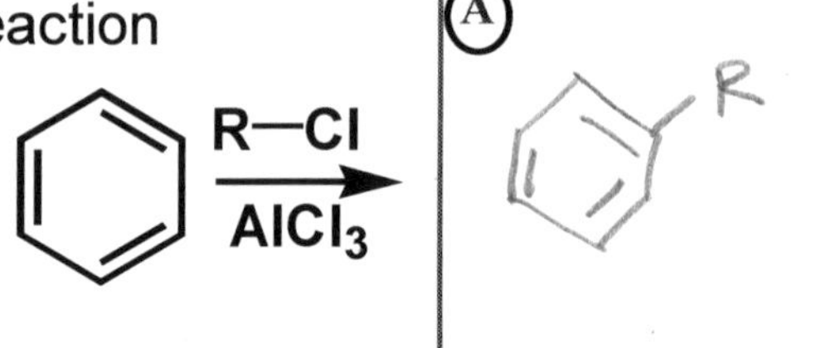

II. How is the electrophile generated?

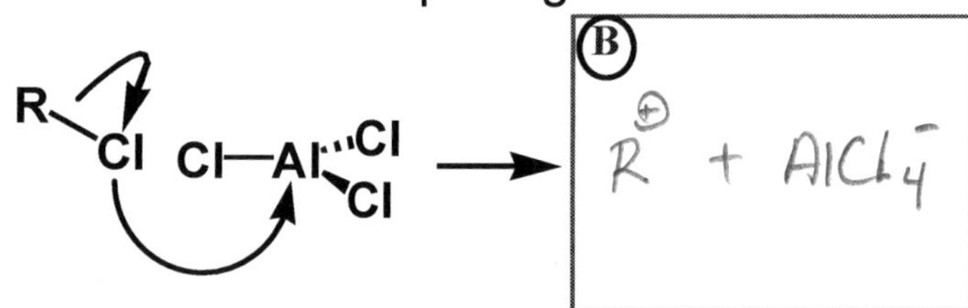

III. The Mechanism:

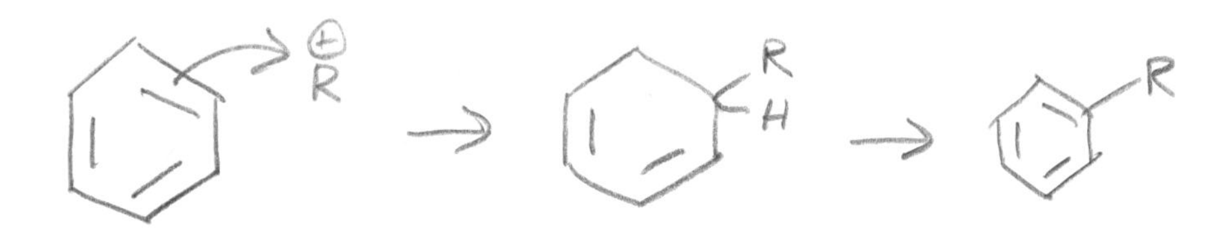

* Carbocation Rearrangement w/ "R" group *

Notes:

Carbocation Rearrangement in F.-C. Alkylation

When looking at any reaction in which a carbocation intermediate is involved, you must be wary of carbocation rearrangement:

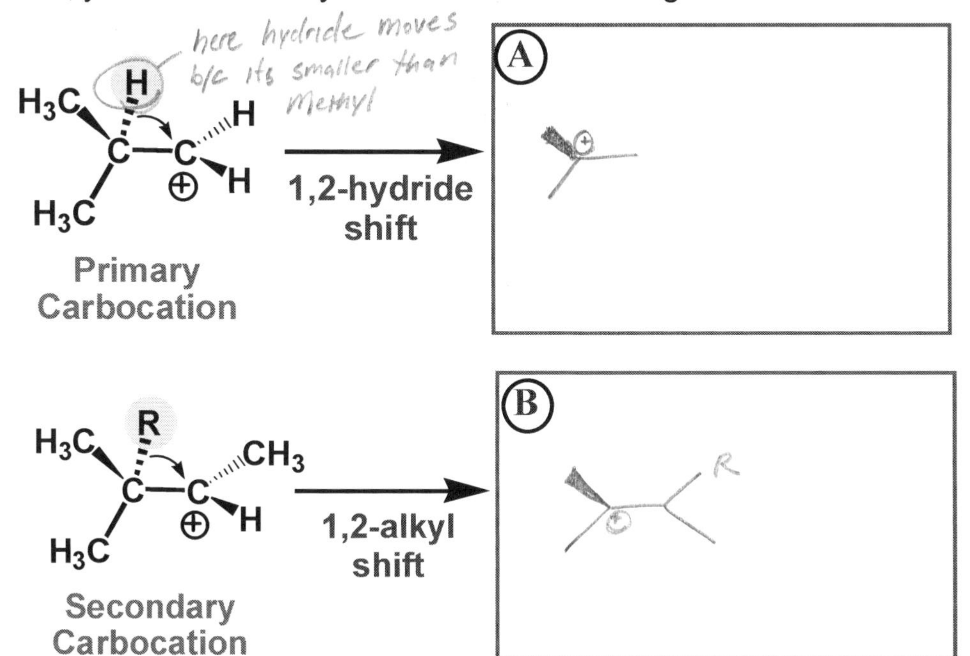

<u>Molecules tend towards the most stable (lowest energy) state***</u>

Notes:

Carbocation Rearrangement in F.-C. Alkylation

Give the major product of the following reaction:

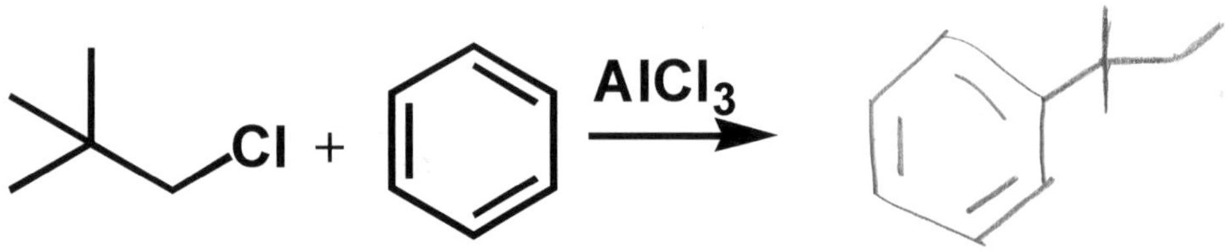

Cl + $AlCl_3$

To solve, go through the mechanism:

1. Make electrophile:

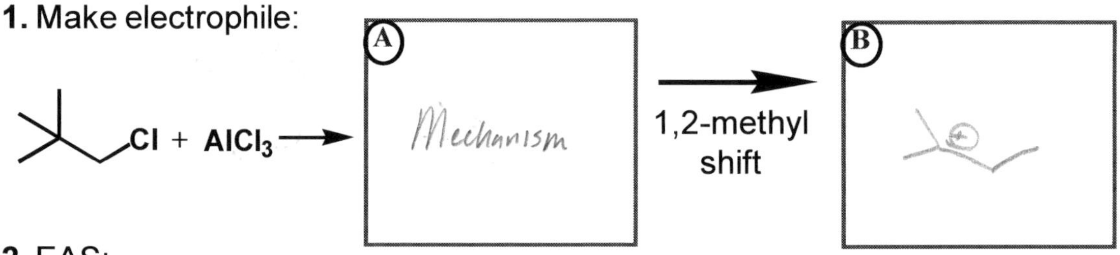

Cl + $AlCl_3$ →

A

Mechanism

1,2-methyl shift

B

2. EAS:

Notes:

Reactions of *Subsituents* on Arenes

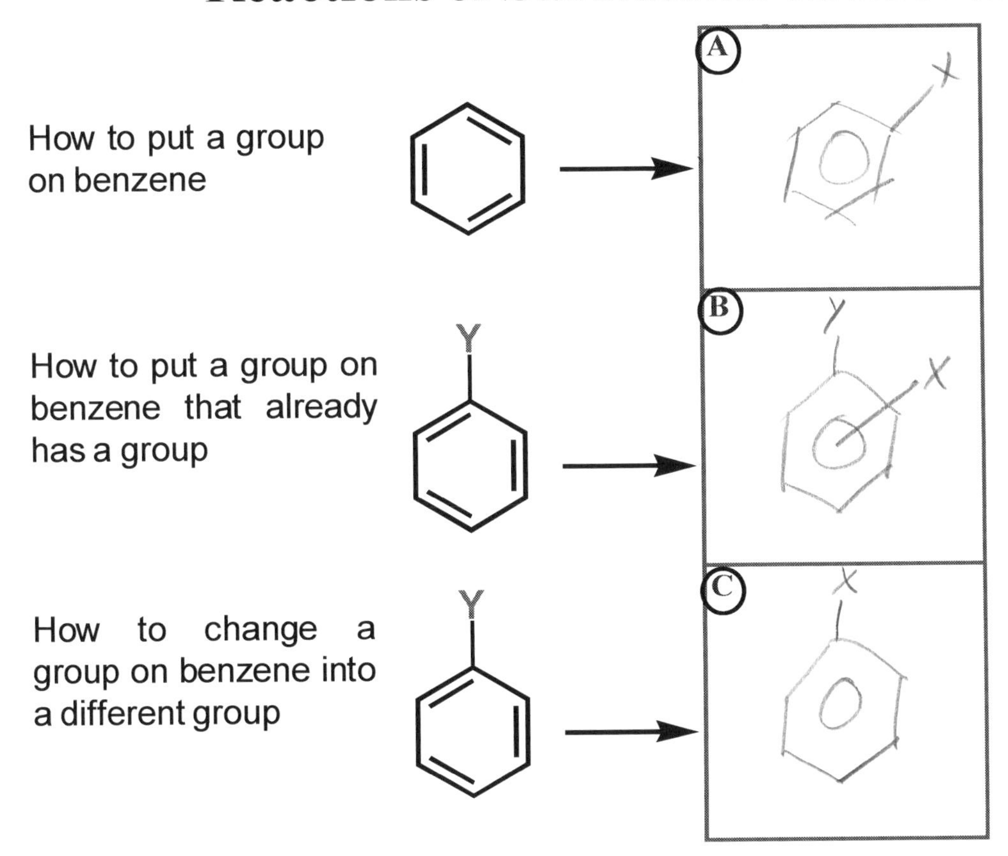
How to put a group on benzene
How to put a group on benzene that already has a group
How to change a group on benzene into a different group
Y
Y
A
B
C
X
Y
X
X

Notes:

Review from Organic 1

Benzylic sites can be brominated (via a radical chain mechanism)

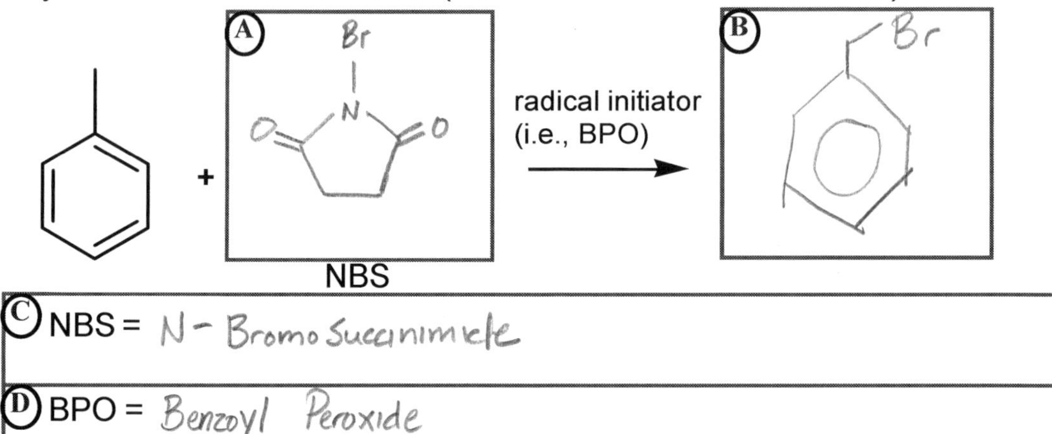

Benzylic bromination is extremely useful because Br is

and consequently can be further changed to other groups …

Notes:

Review: Nucleophilic Substitution

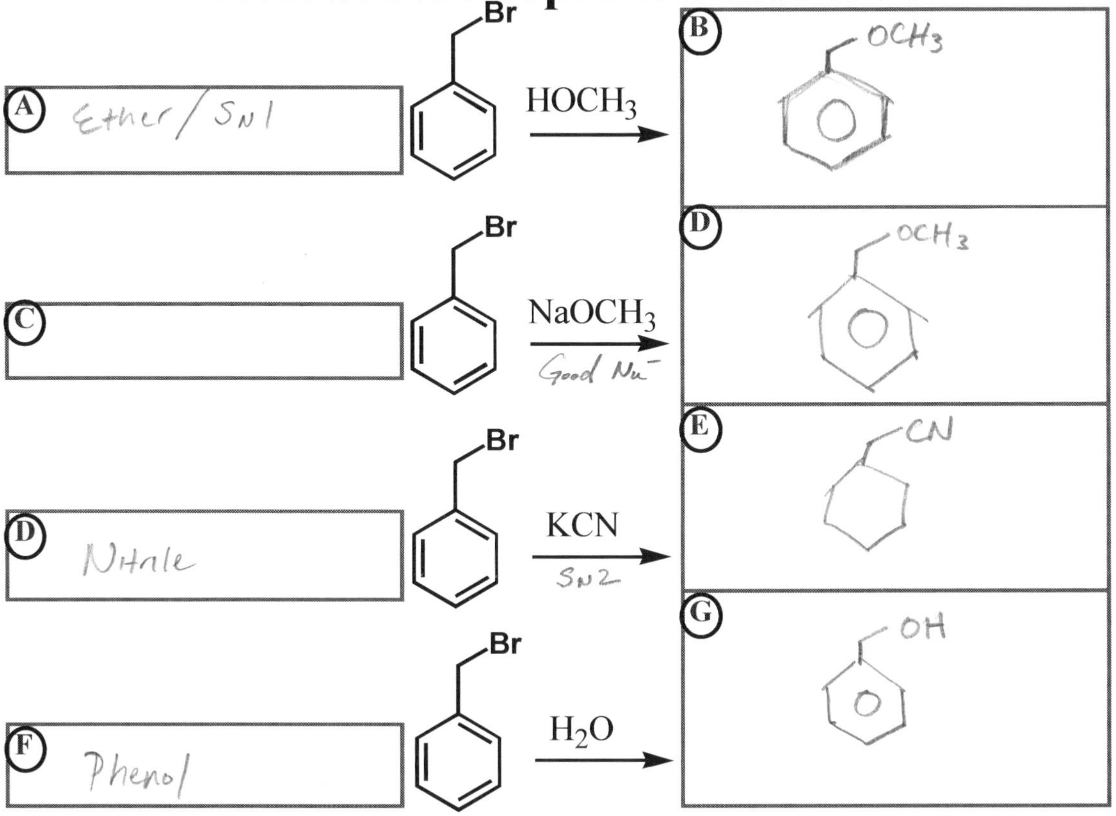

(A) Ether / S_N1

Br — benzyl bromide, $HOCH_3$ →

(B) OCH_3

(C)

Br — benzyl bromide, $NaOCH_3$ →
Good Nu^-

(D) OCH_3

(D) Nitrile

Br — benzyl bromide, KCN →
S_N2

(E) CN

(F) Phenol

Br — benzyl bromide, H_2O →

(G) OH

Notes:

From Lecture Set 1: Organometallic Routes

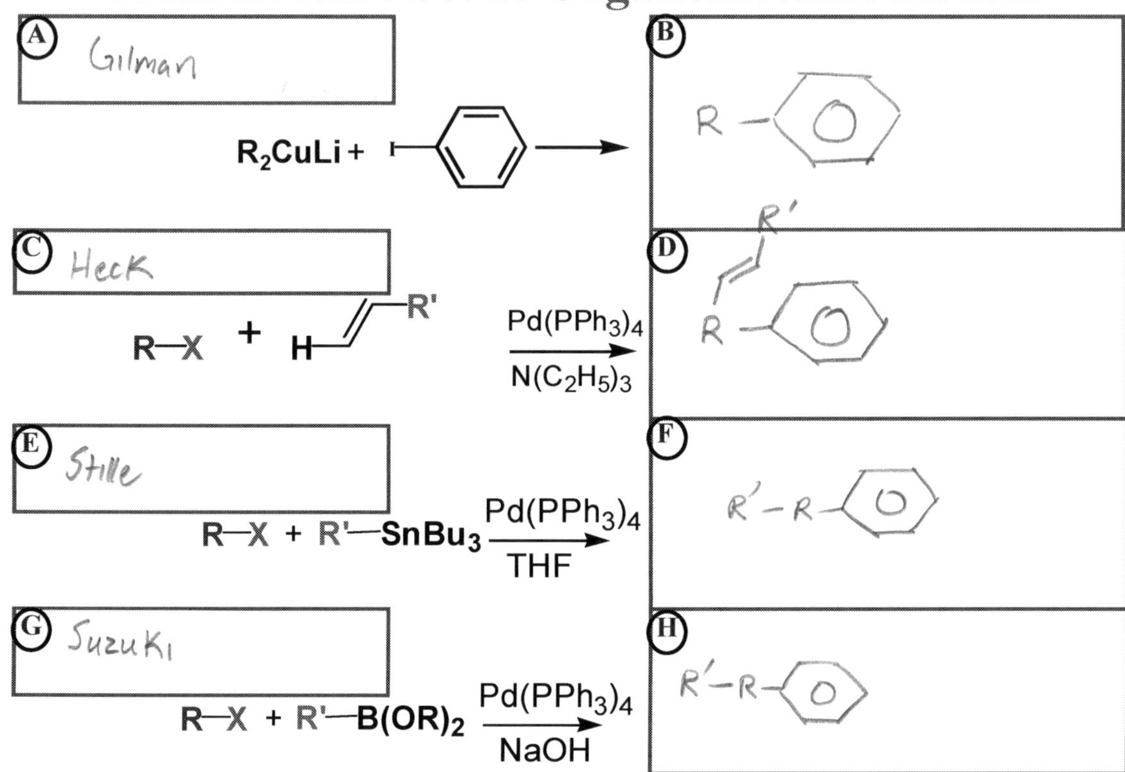
A
Gilman
B
R₂CuLi +
R
C
Heck
D
R—X
+
H
R'
Pd(PPh3)4
N(C2H5)3
E
Stille
F
R—X + R'—SnBu3
Pd(PPh3)4
THF
R'—R
G
Suzuki
H
R—X + R'—B(OR)2
Pd(PPh3)4
NaOH

Notes:

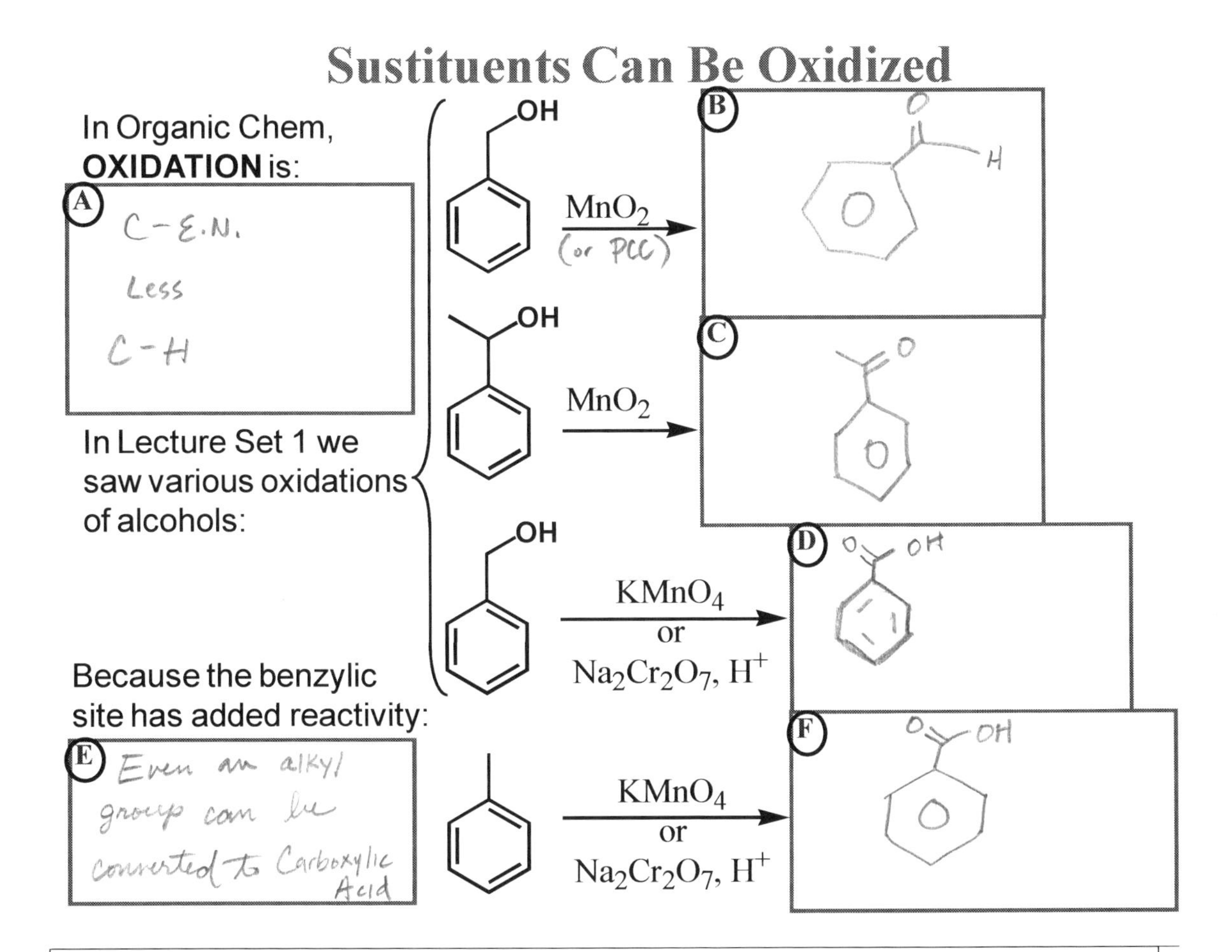
Sustituents Can Be Oxidized
In Organic Chem, OXIDATION is:
A
C-E.N.
Less
C-H
In Lecture Set 1 we saw various oxidations of alcohols:
Because the benzylic site has added reactivity:
E
Even an alkyl group can be converted to Carboxylic Acid
OH
MnO2
(or PCC)
B
O
H
OH
MnO2
C
O
OH
KMnO4
or
Na2Cr2O7, H+
D
O
OH
KMnO4
or
Na2Cr2O7, H+
F
O
OH

Notes:

Sustituents Can Be Reduced

Remember that in organic chemistry, **REDUCTION** is:

(A) more: CH bonds, less C-EN bonds

We've seen such a reaction for an alkene:

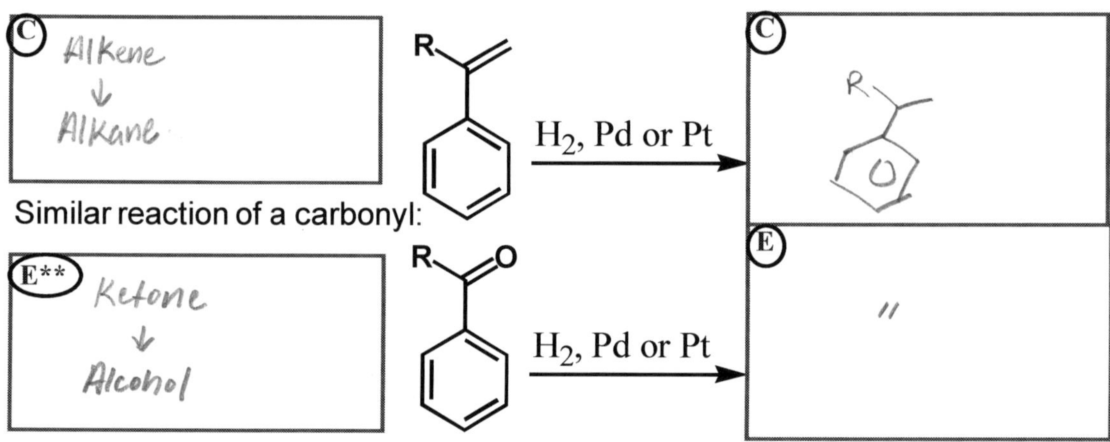

**only works if the C=O is right next to arene

Notes:

Sustituents Can Be Reduced

Such conditions also reduce for triple bonds to nitrogen:

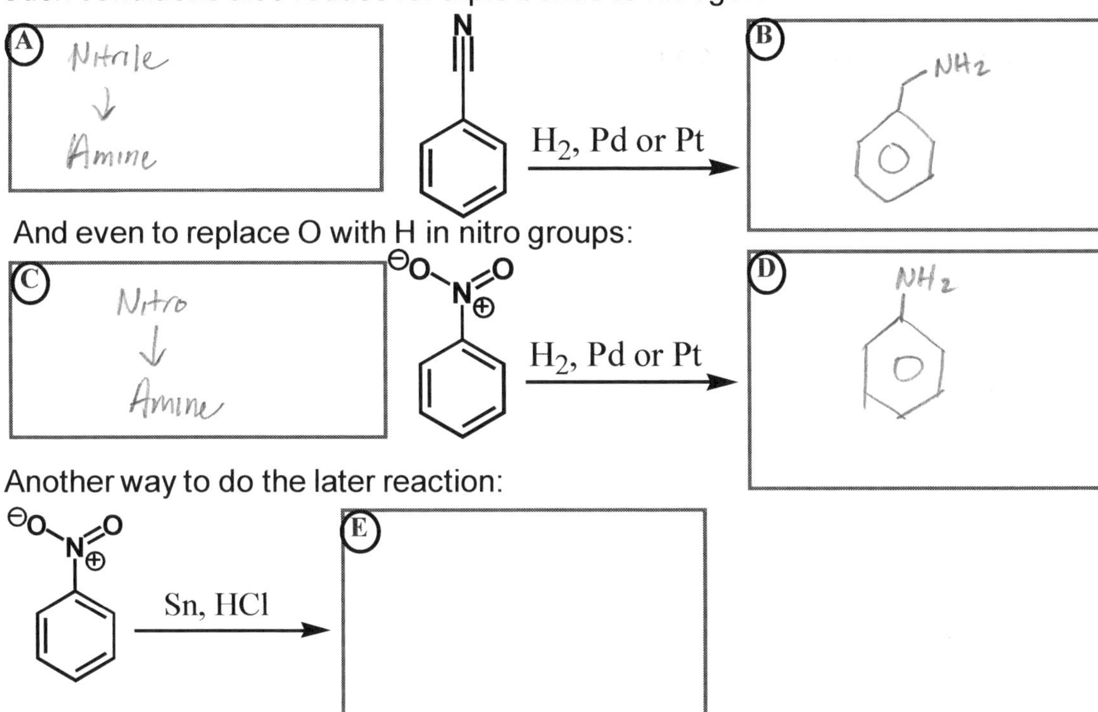

Notes:

Sustituents Can Be Reduced

Two other ways to reduce ketones all the way to alkyl groups:

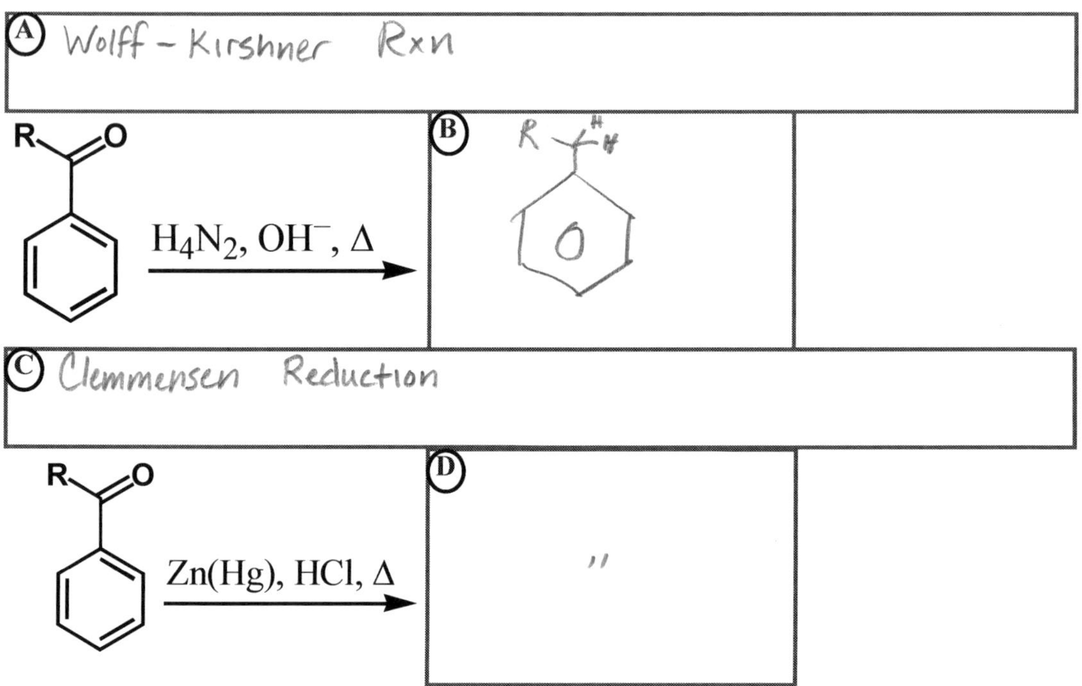

** these work on ketones that are NOT next to benzene as well (unlike H_2/Pd)

Notes:

Set 4 Checklist

❑ Be Able To:

❑ Describe the conditions required for aromaticity and identify a molecule as being aromatic, antiaromatic, or neither.

❑ Determine whether or not a given heterocycle is a good base, and if so which lone pair is most available for interaction with an acid

❑ Name monosubstituted benzene derivatives, including those with common names

❑ Provide arrow-pushing mechanisms for all of the electrophilic aromatic substitution reactions given in this lecture set: **Chlorination, Bromination, Iodination, Nitration, Sulfonation, Friedel-Crafts Acylation,** and **Friedel-Crafts Alkylation**.

❑ Given a reactant and product, provide reagents, or given a reactant and reagents, provide products for the electrophilic aromatic substitution reactions given in this lecture set.

❑ Rank arenes in terms of reactivity to electrophilic aromatic substitution and identify substituents as either **slightly activating**, **activating**, **slightly deactivating**, or **deactivating**.

❑ Use reactions to transform or replace substituents that are already present on an arene:

- **❑ Benzylic bromination**
- **❑ Nucleophilic substitution**
- **❑ Alkylation using a Gilman reagent**
- **❑ Pd-catalyzed coupling (Heck, Stille, Suzuki) to add an aryl or vinyl group**
- **❑ Oxidations**
- **❑ Reductions (including Clemmensen and Wolf-Kishner)**

❑ Do everything listed in previous checklists

❑ Do all problems at the end of the chapter in your text.

Notes:

Lecture Set 5:
Substituted Benzene:
Nomenclature and Reactions

Suggested Reading:

Suggested Problems:

Nomenclature of Disubstituted Benzene

In Lecture Set 4, we focused on naming monosubstituted benzene derivatives and making monosubstituted benzene derivatives via EAS on benzene. Now we will look at the nomenclature and preparation of disubstituted and polysubstituted arenes.

When naming benzene derivatives, you can use benzene as the parent chain and numbers to denote positions of substituents, like we learned for cycloalkanes in Organic 1. However, there is another widely-used nomenclature method for disubstituted benzenes you must also know.

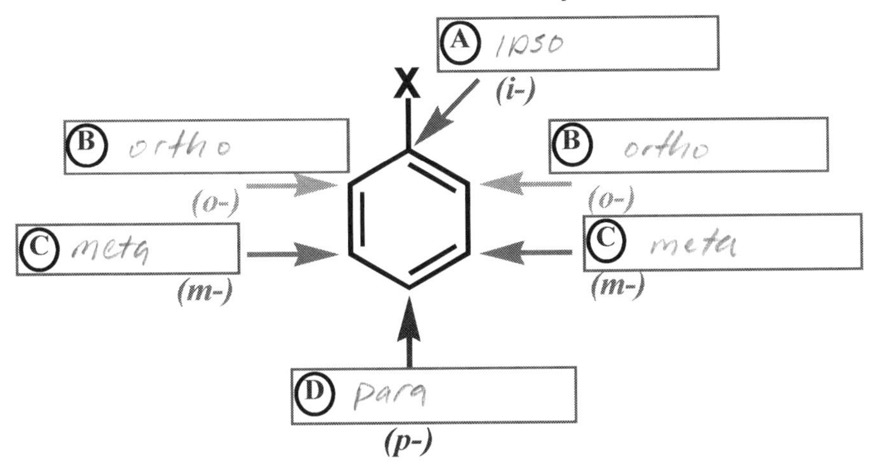

Notes:

Some Examples

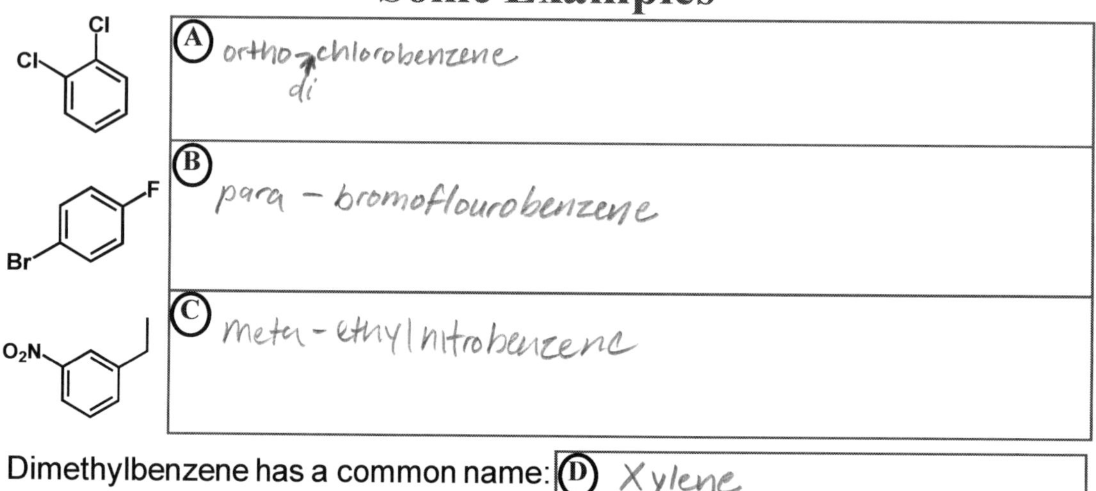

(A) ortho-chlorobenzene

Dimethylbenzene has a common name: (D) Xylene

(B) para — bromoflourobenzene

(C) meta - ethyl nitrobenzene

(E) ortho-xylene

NOT 1,2-dimethylbenzene

(F) meta-xylene

(G) para-xylene

Notes:

More Examples

If a particular molecule contains a benzene derivative with a common name, then use that as the parent, and the substituent that is part of the parent structure always is given the number 1...

Notes:

Nomenclature of Disubstituted Benzene

Example: Name the following molecules using both IUPAC and common nomenclature methods:

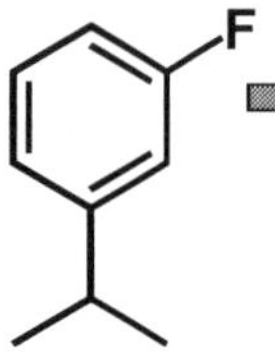

F

(A)

(B)

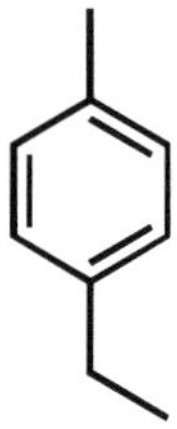

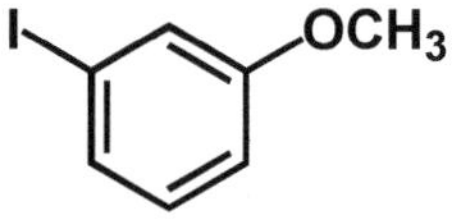

OCH_3 I

(C)

(D)

CN

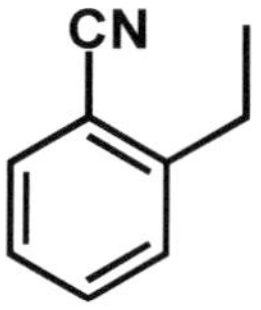

Notes:

Nomenclature of Polysubstituted Benzene

If there are more than two substituents on the benzene ring, you must use IUPAC (numbering) notation.

Example: Name the following molecules using both IUPAC and common nomenclature methods:

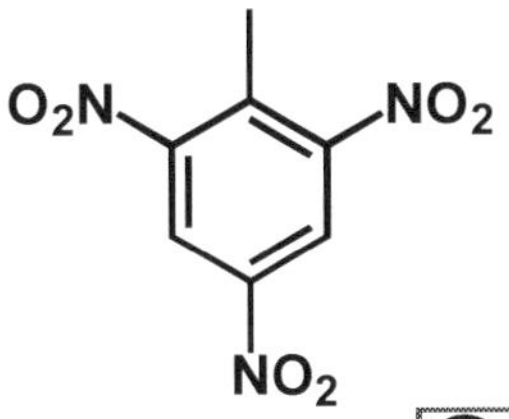

O_2N NO_2 NO_2

(A)

(B)

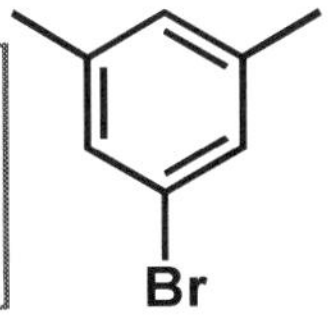

Br

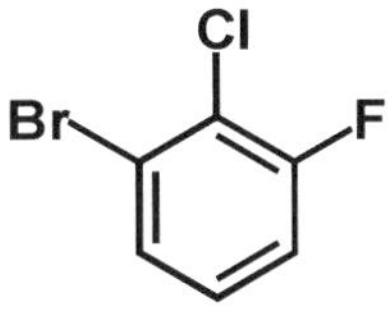

Cl

Br F

(C)

Notes:

EAS on Substituted Benzene

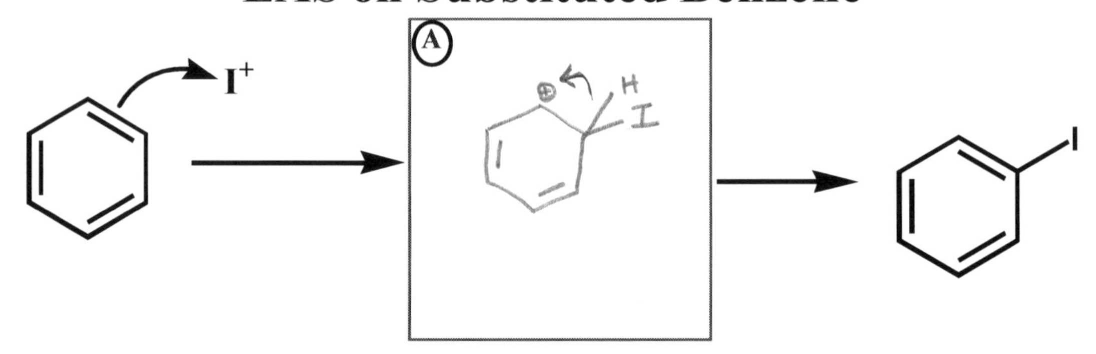

How will this be effected if there is a substituent on benzene?

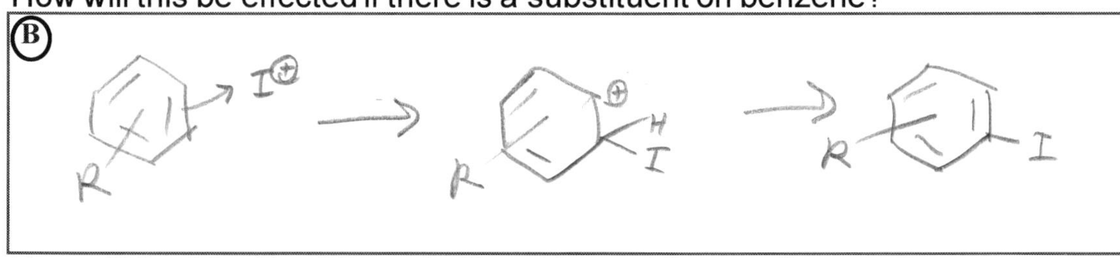

How will the identity of the substituent impact the reaction rate?
How will the identity of the substituent impact the identity of the product?

KEY Intermediate: (C)

Notes:

Tuning the Reactivity of Arenes

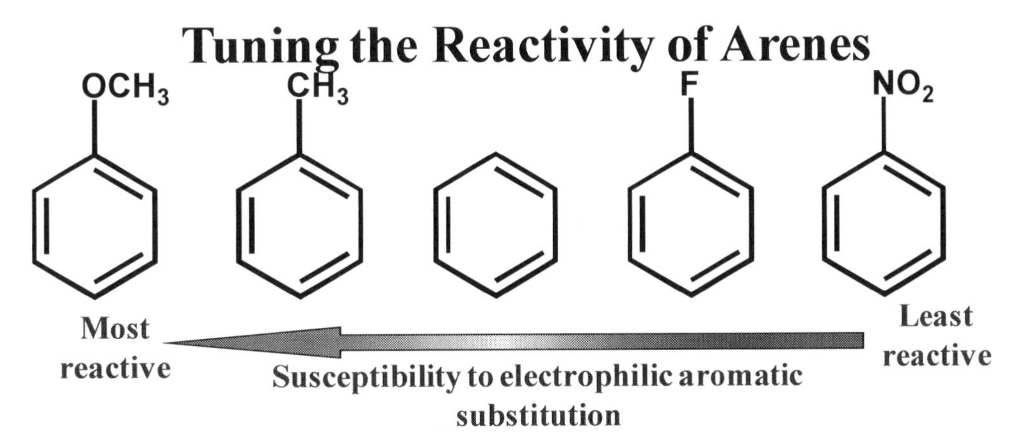

We have a carbocation intermediate, so a more stable carbocation will lead to its more rapid formation.

If we put **electron donor groups** on the benzene ring, it is

(A) more reactive than unsubstituted benzene.

If we put **electron withdrawing groups** on, the arene will be

(B) less reactive than unsubstituted benzene.

This is a lot like our acid-base rules when we wanted to stabilize an anion. Now we want to stabilize a cation (opposite rules!).

Notes:

:X—Y

Here are some guiding principles for predicting relative activating/deactivating potential of substituents

If the element (X) **DIRECTLY** attached to the benzene ring has a **lone pair**,

(A) Donates e^- by resonance → substituents are activating

Except **HALOGENS**:

(B) (X E.N.) slightly deactivating

2. If the substituent is a **HYDROCARBON** (i.e., alkyl, vinyl, or aryl group):

(C) slightly activating

3. If an element (Y) **ADJACENT** to the directly attached atom is more **electronegative** than carbon,

(C) (e^- w/d) most deactivating subs.

If the substituent has a positive charge next to benzene:

(D) Deactivate

If the substituent has a negative charge next to benzene:

(E) Activate

Notes:

Substituent Effect on Reactivity

So now we've seen the effect of substituents on the names **and** reactivity of aromatic compounds. Recall the trends we observed for substituents on benzene as to the reactivity of the ring in electrophilic aromatic substitution.
In this class, we place substituents into **four general EAS Reactivity Groups**:

(A) Activating O–Ẍ–Y (X ≠ halogen)
(B) Slightly Activating (hydrocarbons)
(C) Slightly Deactivating (halogens)
(D) Deactivating O–Ẍ–Y (Y = EN)

← Benzene = 0

Notes:

What effects the ability of an element to donate electrons to carbon?

Size – if the element is not in the same row as carbon, it is too big to effectively overlap the pi system.

•Electronegativity – because fluorine is so electronegative, it does not donate electrons to the pi system enough to activate the molecule, even though it is the same size.

Combining the rules with these 2 guiding principles, we can group many substituents as activating or deactivating (relative to benzene)

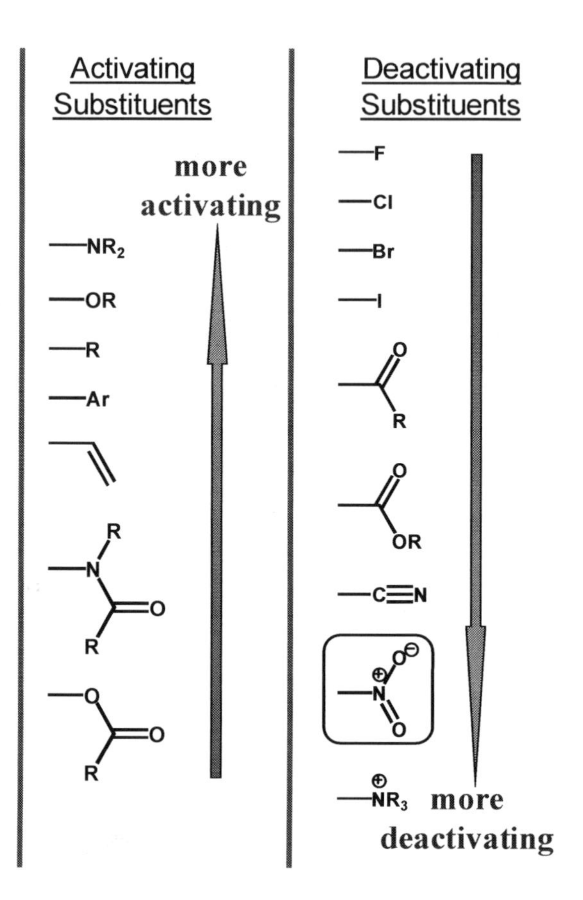

Notes:

Reactivity of Arenes Problem

Example. Classify each of the given substituents as being Slightly Activating (SA), Activating (A), Slightly Deactivating (SD), or Deactivating (D), relative to H (plain benzene), with respect to reactivity in Electrophilic aromatic substitution.

(i) CH_3 — SA
(ii) O — A
(iii) SO_3H — D
(iv) N — A
(v) SA
(vi) Cl — SD
(vii) ⊕N — D
(viii) NO_2 (more EN) — D
(ix) Br — SD
(x) O (more EN) — D

Notes:

Tuning the Reactivity of Arenes Problem

Example. Rank these arenes from 1-5 in terms of their reactivity towards electrophilic aromatic substitution, 1 being most reactive and 5 being least reactive. Explain your selections.

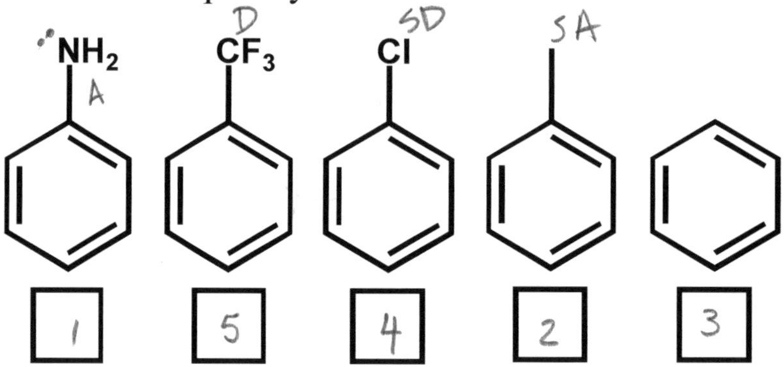

Notes:

Substituent Effects on EAS Substitution Site

If there is already a substituent on a benzene ring and we then attempt to further substitute it, we need to consider what effects the substituent has not only on its reactivity, but also on the distribution of products (i.e., what is the major product?). If a monosubstituted arene is subjected to EAS, we can get *o*-, *m*-, and *p*- products:

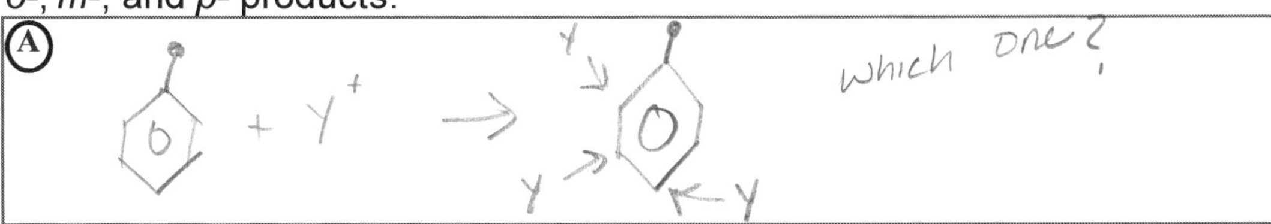

In order to understand which product(s) are formed in highest yield, we must examine the mechanism, keeping in mind one of the ***generally applicable principles*** used in examining organic chemical transformations:

B which Intermediate is most stable?

*EAS = Electrophilic aromatic substitution

Notes:

All Activating Substituents

Anisole: *ortho*, *meta*, or *para*?

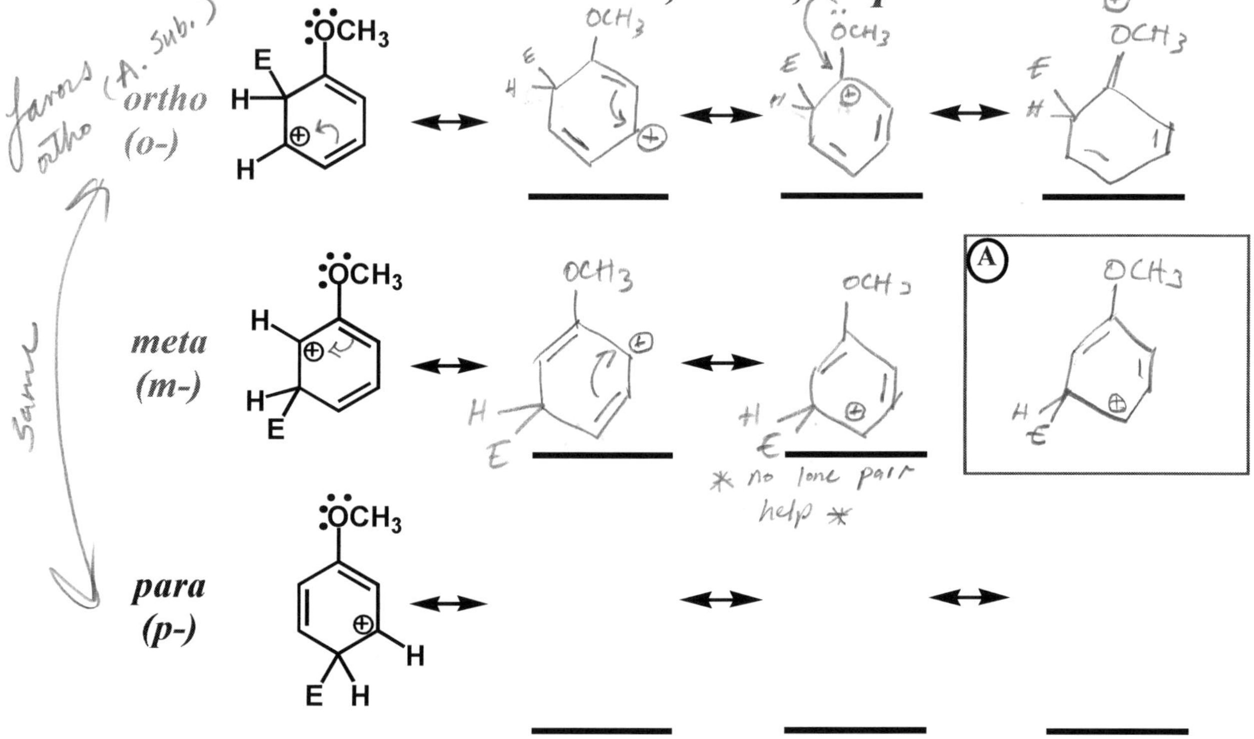

favors ortho (A. Sub.)

ortho (*o-*)

:ÖCH₃ E H H ⊕ ↔ OCH₃ E H ⊕ ↔ ÖCH₃ E H ⊕ ↔ ⊕ OCH₃ E H

same

meta (*m-*)

:ÖCH₃ H ⊕ H E ↔ OCH₃ ⊕ H E ↔ OCH₃ H E ⊕

* no lone pair help *

Ⓐ OCH₃ H E ⊕

para (*p-*)

:ÖCH₃ ⊕ H E H ↔ ↔ ↔

Notes:

Toluene: *ortho*, *meta*, or *para*?

For carbocations, recall stability is greatest for 3 (> 2 > 1) due to hyperconjugation

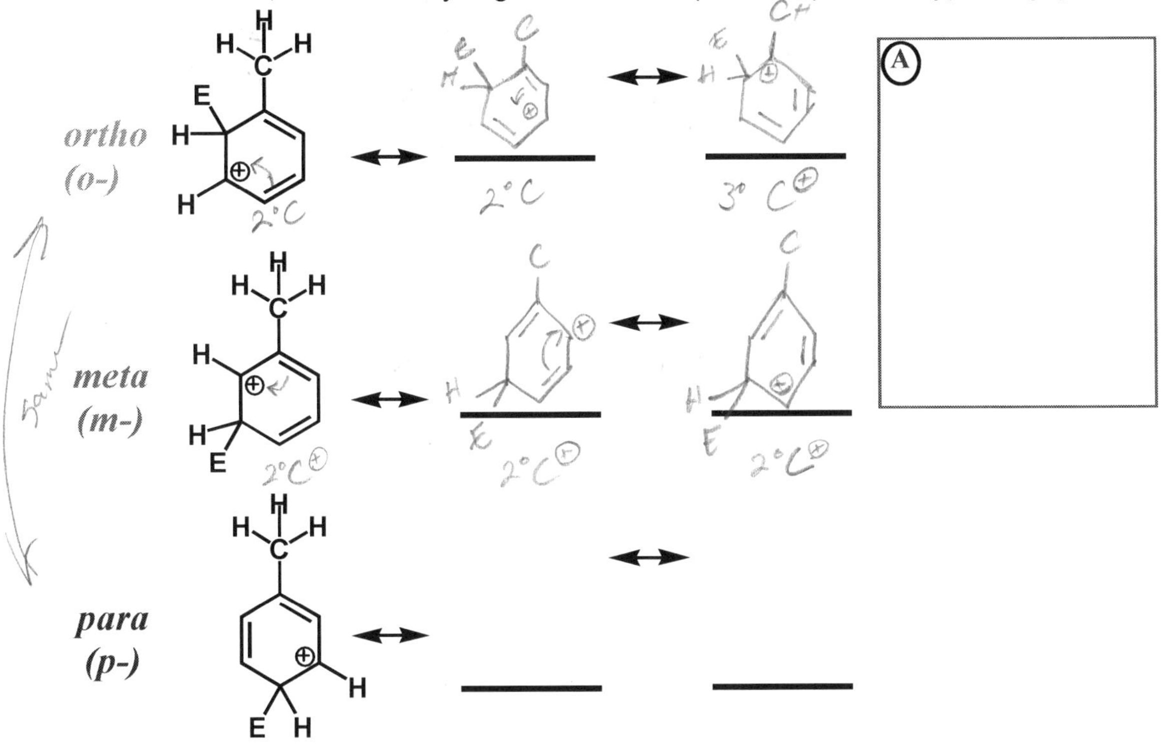

ortho (*o-*)

meta (*m-*)

para (*p-*)

A

Notes:

Deactivating Sub.

Benzaldehyde: *ortho*, *meta*, or *para*?

ortho (*o-*)

meta (*m-*)

para (*p-*)

BAD Destabilizing b/c + next to δ+

no ⊕ next to δ+

Ⓐ Activating — ortho & para

SA - ortho + para

SD - ortho + para

Deactivating — ONLY meta

Notes:

Substituent Effects on Product Orientation in EAS*

A general trend worth noting is that we can correlate a substituent's **EAS Reactivity Group** with its **EAS Directing Type**:

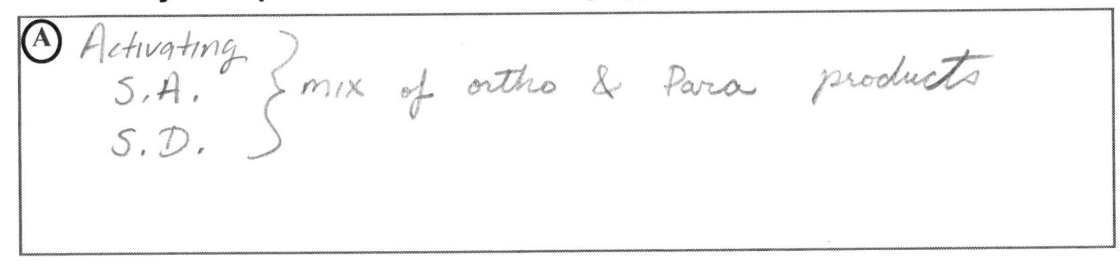

and

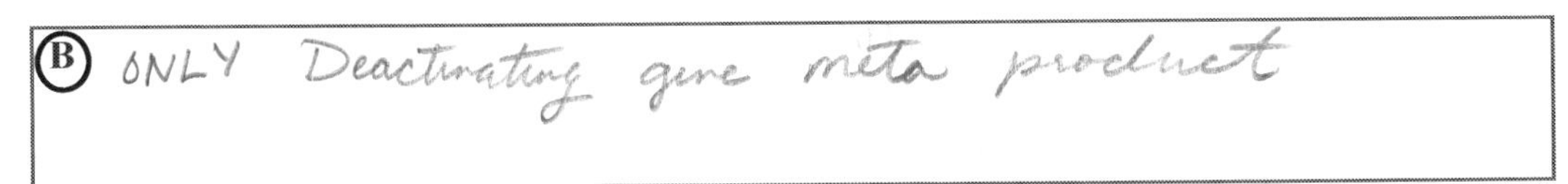

(B) ONLY Deactivating give meta product

Remember that these are *general trends* to help us predict the **major** product; you typically won't get ONLY the favored product(s)

*EAS = Electrophilic aromatic substitution

Notes:

Problem

Example: Give the major product(s) of the following reactions using your knowledge of whether the existing substituent is o-/p- directing or m-directing.

Deactivating so meta position addition

(A) $\xrightarrow[FeBr_3]{Br_2}$ adds Br to Ring — Br

Activating so o/p addition

(B) $\xrightarrow[H_2SO_4]{HNO_3}$ adds NO_2 — Sub. NO_2 + Sub. NO_2

(C) Br SD, R–C(=O)–Cl $\xrightarrow{AlCl_3}$ adds acyl group — Br, O, R + Br, O, R

Notes:

The *ortho:para* ratio

A complication is that the activating and slightly deactivating substituents direct substitution at both *ortho* and *para* positions. It would be useful to be able to predict how much o- and how much p- product will be formed. One consideration is:

Ⓐ 2 ortho sites but 1 para site so 2:1 ratio

Another point to consider is:

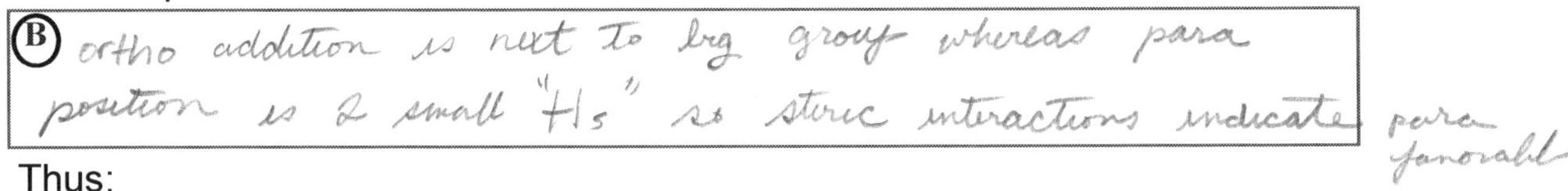

Ⓑ ortho addition is next to lrg group whereas para
position is 2 small "H's" so steric interactions indicate para
favorable

Thus:

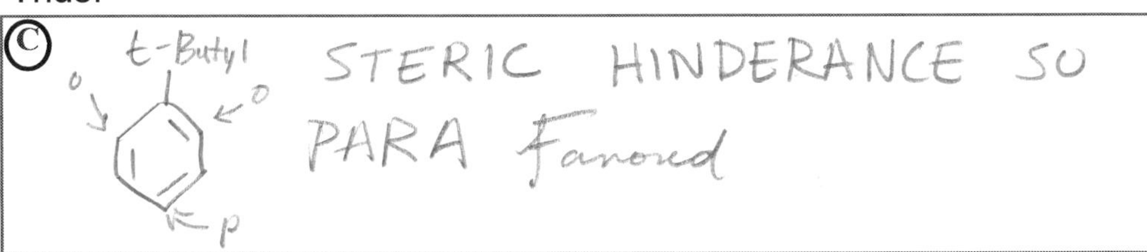

Otherwise, it is somewhat difficult to predict intuitively what the product distribution will be in a real-world context.

Notes:

What if there are multiple substituents?

Often, a starting arene will have two substituents on it. In these cases, you need to consider the directing preference of both substituents. As a general rule:

(A) More Activating Sub. dictates rxn product

You also must consider:

(B) putting E (added group) in b/w 2 sub. is difficult or unfavorable

Examples:

OCH_3 (A o/p), Cl (SD, para) $\xrightarrow[FeBr_3]{Br_2}$ (C) Br, OCH_3, Cl (← or Br here)

(D) OCH_3 dominates

SA, large SA $\xrightarrow[H_2SO_4]{HNO_3}$ (E) NO_2, t-But.

(F) STERIC HINDERANCE EFFECTS

Notes:

Problem

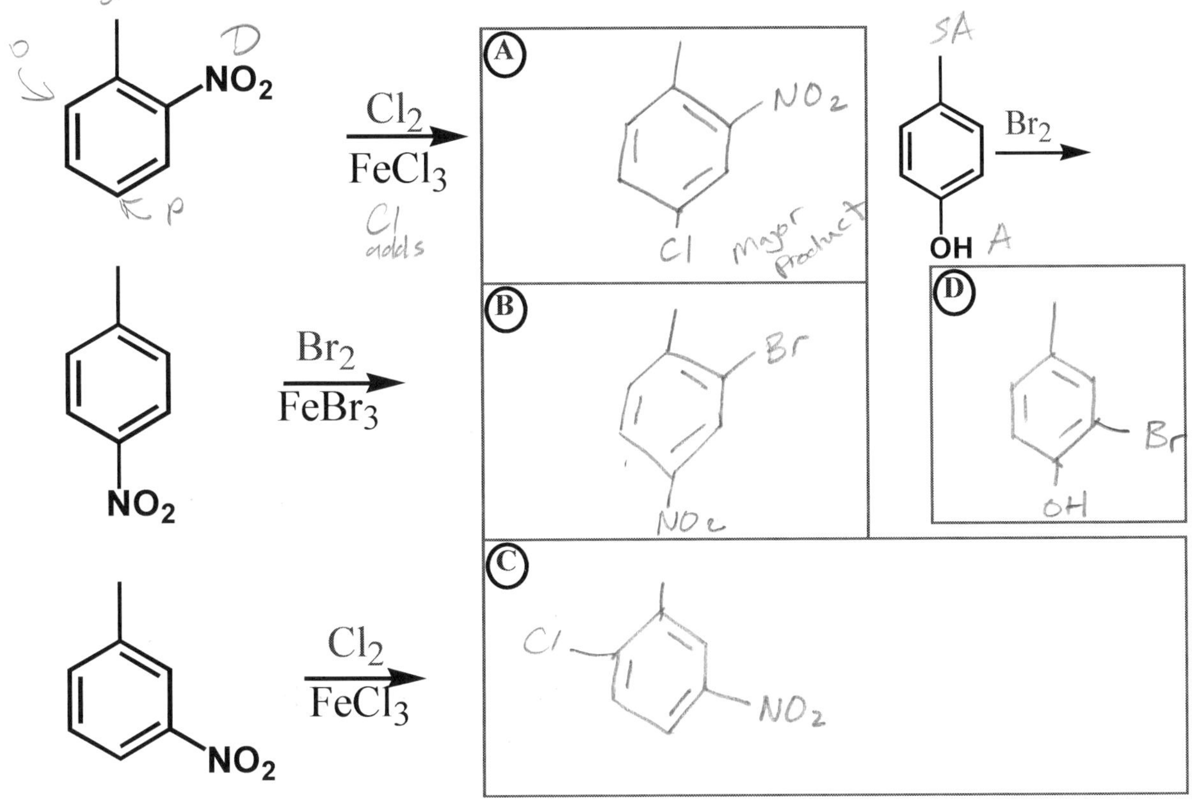

SA

o

D

NO_2

p

$\xrightarrow[FeCl_3]{Cl_2}$

Cl adds

(A)

NO_2

Cl major product

SA

$\xrightarrow{Br_2}$

OH A

NO_2

$\xrightarrow[FeBr_3]{Br_2}$

(B)

Br

NO_2

(D)

Br

OH

NO_2

$\xrightarrow[FeCl_3]{Cl_2}$

(C)

Cl

NO_2

Notes:

Multistep Synthesis/Retrosynthetic Analysis

Example: Give the best synthesis of the following targets from benzene in as many steps as necessary, using the reactions presented in the notes thus far.

I. Some simpler ones:

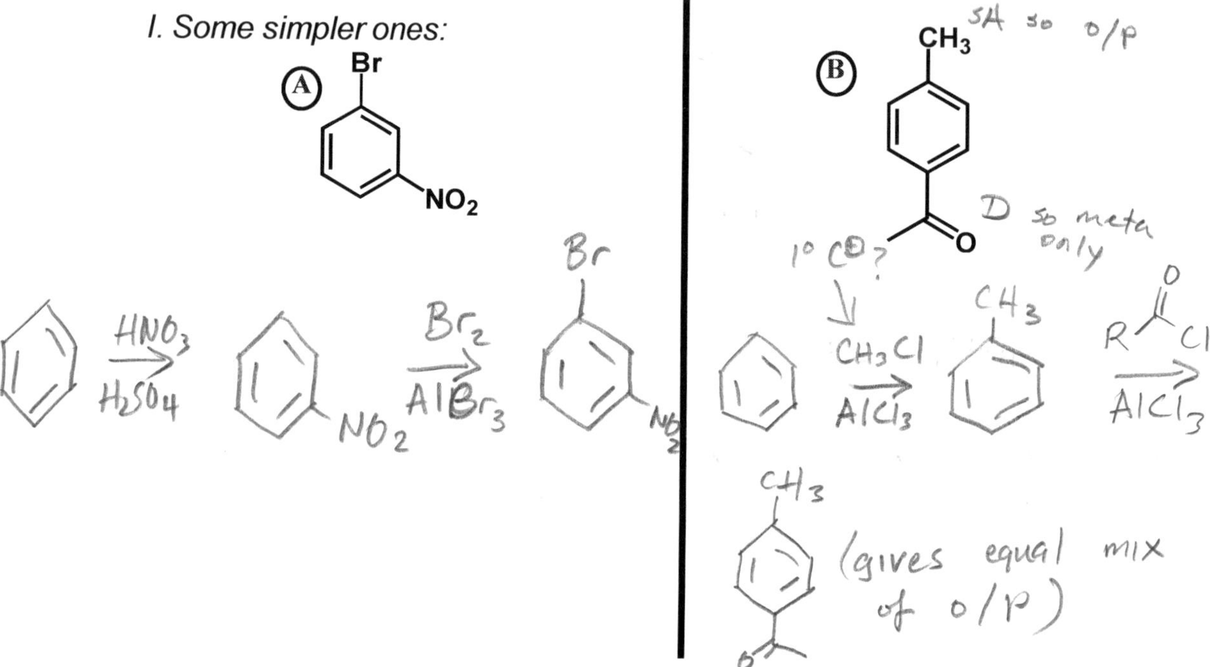

**The 1st thing to ask is "which substituent goes on last???

Notes:

1) Br goes on last b/c of meta-placement via NO_2 (if NO_2 goes last Br will add it to o/p position)

Cont'd

II. Some that require substitution followed by reaction of the functional groups:

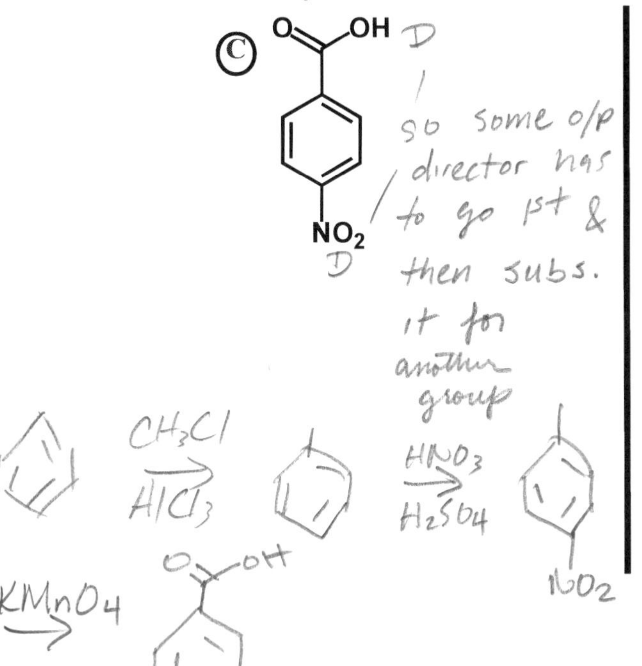

D

NH_2 A so meta

Br SD so o/p

Br

Br_2

$FeBr_3$

HNO_3

H_2SO_4

Br

NO_2

H_2

R. Ni

Br

NH_2

Notes:

Diazonium Salts

Diazonium salts ($R\text{-}N_2^+$)are good precursors for transformation into other functional groups on arenes via the following general reaction scheme:

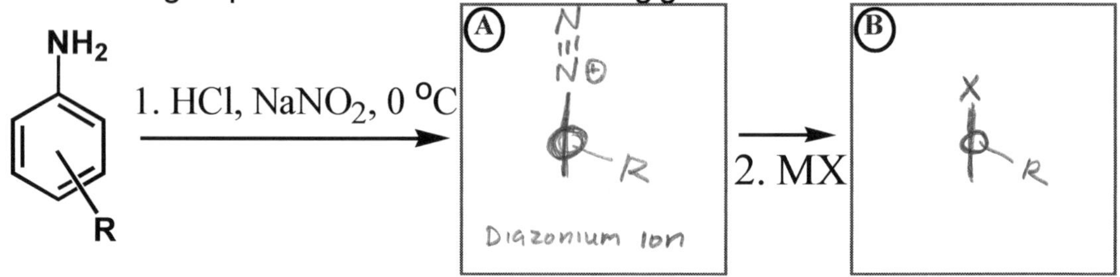

The identity of the group that ends up on the ring depends on what compound MX is added in the second step:

NaI also

$$MX = CuCl, CuBr, CuCN, Cu_2O, KI, HBF_4$$

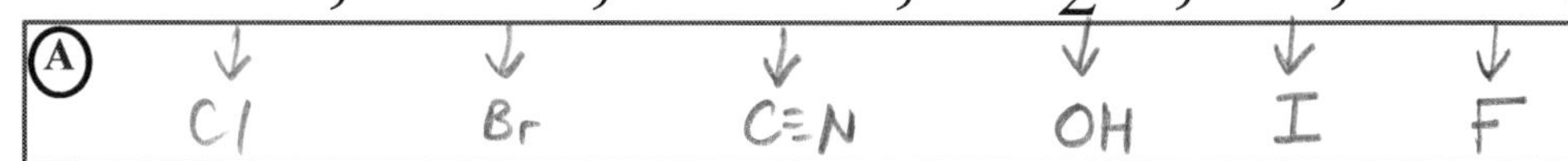

Notes:

Diazonium Salts in Synthesis

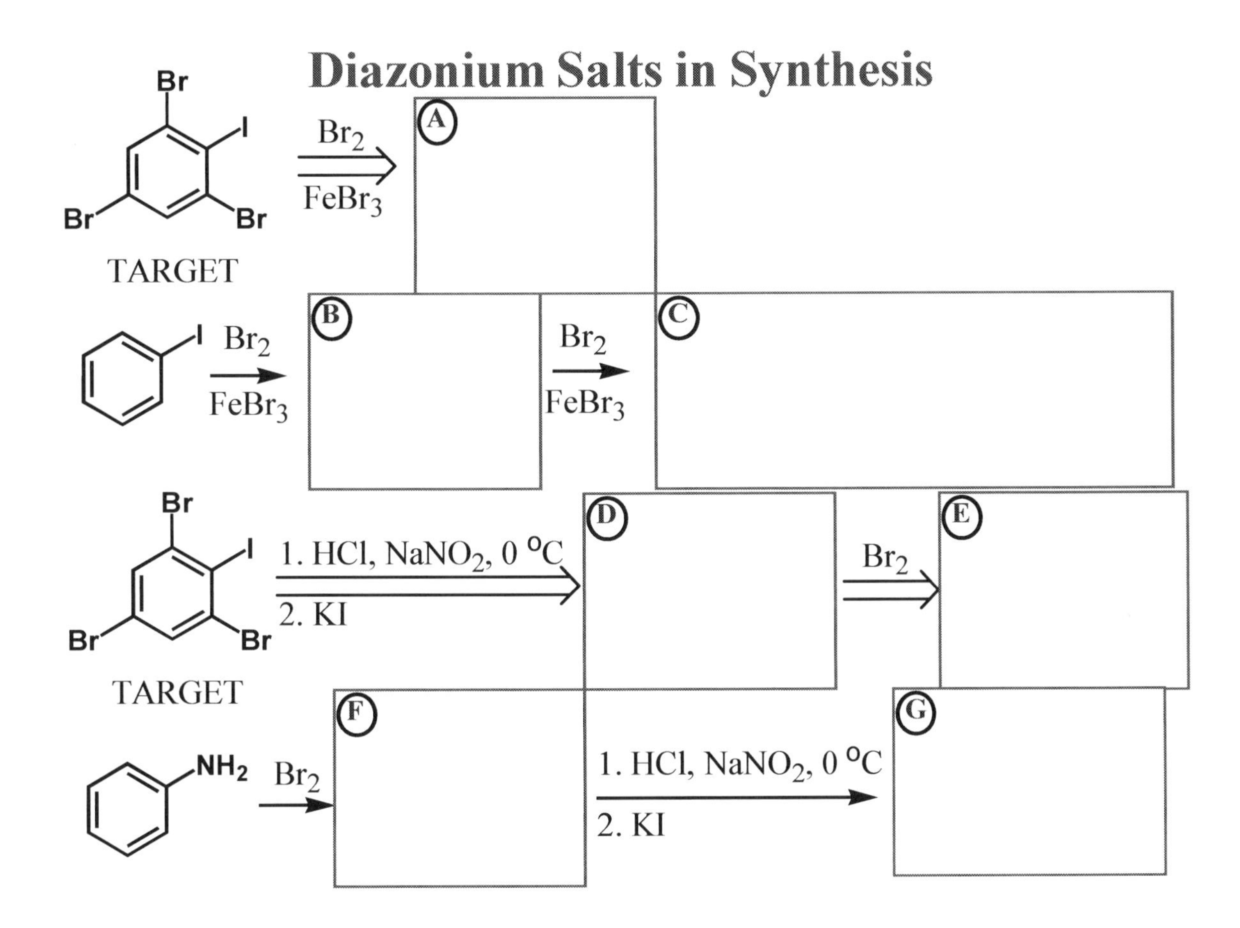

Notes:

Start from Benzene

<u>Nucleophilic</u> Aromatic Substitution I: S_NAr

In Lecture Set 14, we looked at Electrophilic Aromatic Substitution reactions of Arenes. We subsequently explored how substitution of the arene affects the rate and regiochemistry of electrophilic aromatic substitution reactions. Now we will briefly explore how aromatic compounds react with **Nucleophiles** instead of electrophiles. Arenes may react with nucleophiles via:

(A) **S_NAr**

Nucleophilic Aromatic Sub.

Many of the reactivity trends, etc. that we learned for EAS reactions can be used to help understand **S_NAr** reactions; but we have to reverse some of the trends in light of the fact that we are now reacting the aromatic with a nucleophile instead of an electrophile.

It is important to note that, without extreme conditions:

(B) Very limited! ONLY works on Benzene w/ strong e^- w/d groups like NO_2

Notes:

<u>Nucleophilic</u> Aromatic Substitution I: S_NAr

The net S_NAr reaction is replacement of the substituent that is *ortho-* or *para-* to the strongly withdrawing substituent with the nucleophile:

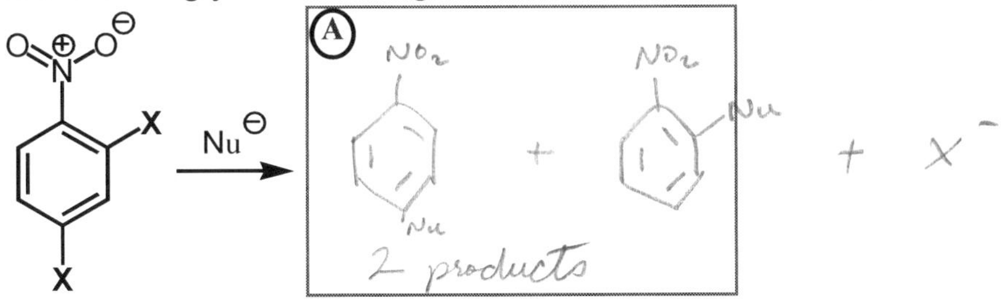

Remember that **for EAS** $-NO_2$ is *m*-directing, but **in S_NAr**:

(B) o/p products (almost always ortho & para)

The S_NAr reaction typically only works when:

(C) Nu⁻ that used to attack must be a stronger base then X

Notes:

<u>Nucleophilic</u> Aromatic Substitution I: S_NAr

Though the name "**S_NAr**" reminds us of S_N2, the mechanism is different (otherwise it wouldn't have its own name!).

Recall the S_N2 mechanism:

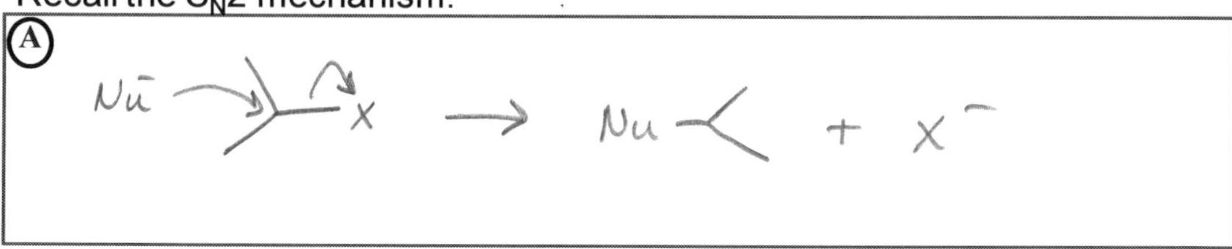

This is not possible when the leaving group is on an aromatic ring (or any sp^2-hybridized atom):

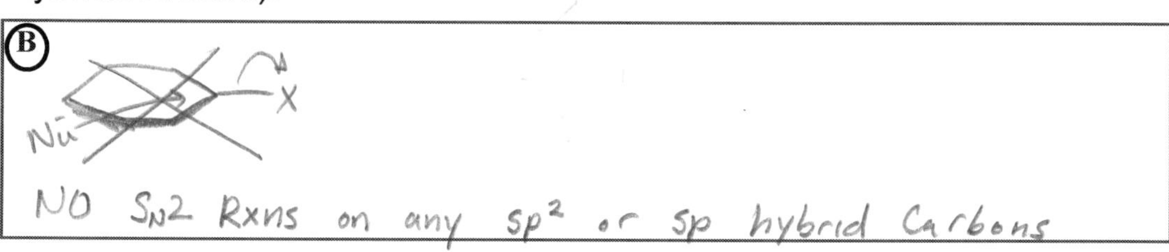

B
Nu⁻
X
NO SN2 Rxns on any sp2 or sp hybrid Carbons

Notes:

The mechanism of S_NAr is thought to be:

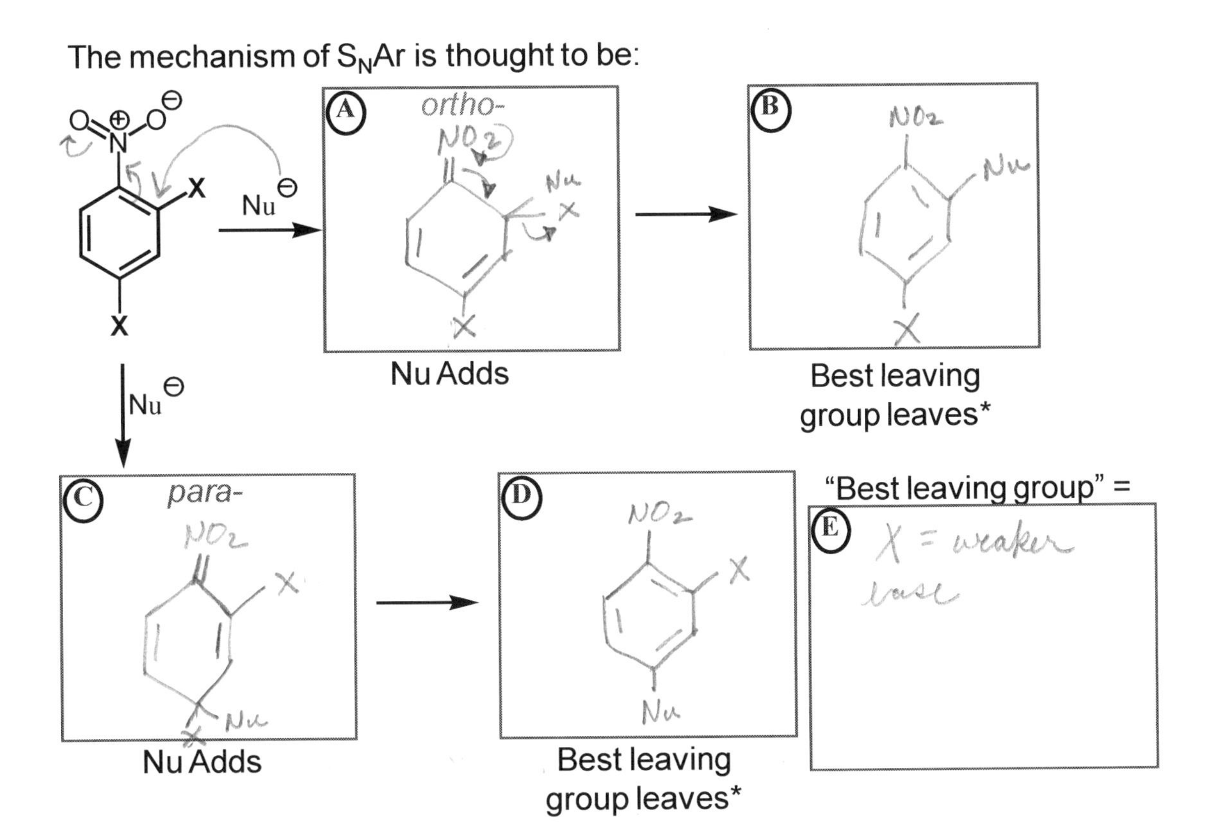

Notes:

S_NAr Examples

NO_2, Cl, Cl (2,4-dichloronitrobenzene) $\xrightarrow{HO^{\ominus}}$

A

NO_2, OH, Cl

or

NO_2, Cl, OH

NO_2, Br (LG) $\xrightarrow{CH_3\ddot{N}H_2}$

B

NO_2

$CH_3-\overset{\oplus}{N}-H$

H

NO_2, CH_3 (Not a Good LG) $\xrightarrow{CH_3O^{\ominus}}$

C

CH_3^- is much less stable anion than CH_3O^- so NO RXN

Notes:

Lecture Set 6: Introduction to Carbonyls and Nucleophilic Acyl Substitution Reactions

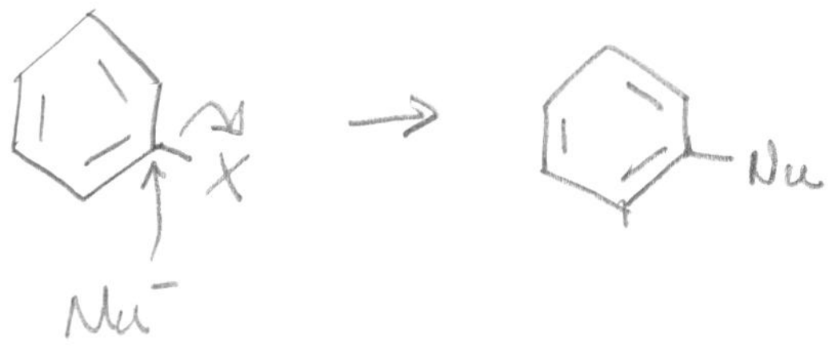

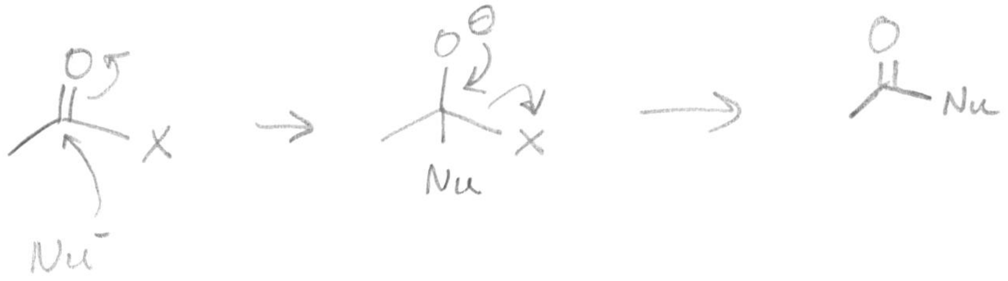

Nu⁻ X Nu

Nu⁻ O X Nu O⁻ X Nu O Nu

Suggested Reading:

Suggested Problems:

Some General Trends in Carbonyl Reactivity

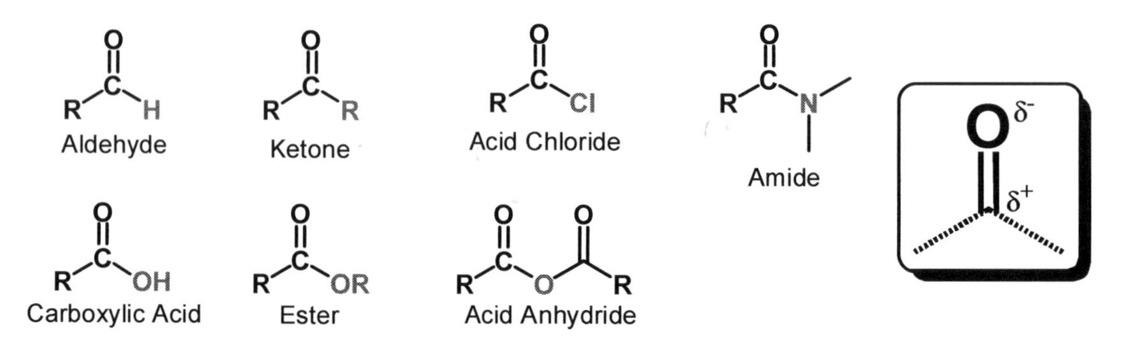

The chemistry of carbonyls is often dominated by the polarity of the C=O bond. Specifically, the partial positive charge on the carbonyl C makes it a potential site for nucleophilic attack:

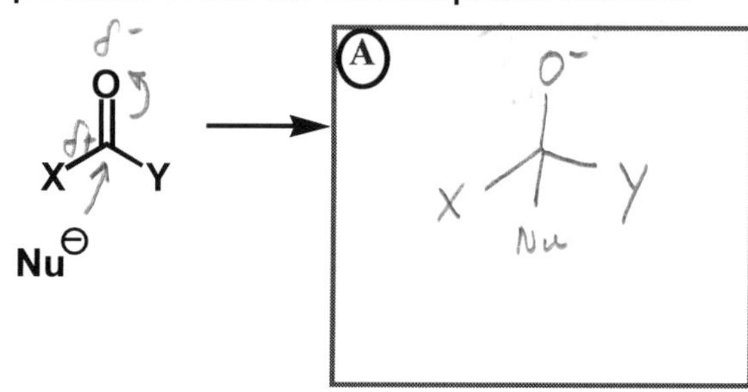

Now what happens downstream of that initial nucleophilic attack is often the defining feature that discriminates the eventual products ...

Notes:

Some General Trends in Carbonyl Reactivity

Consider what could happen to the product (box A) on the previous slide:

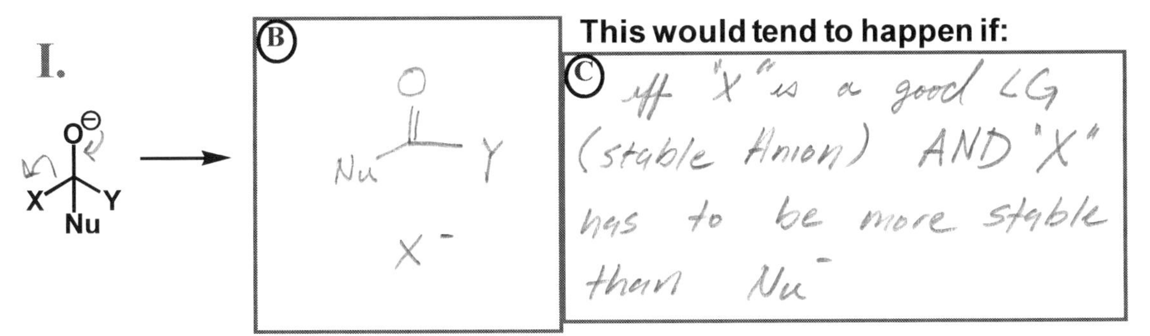

II.

D

OH

X

Y

Nu

H^+

This would tend to happen if:

E

None of LG (X, Y, or Nu) are stable anions (ie. Strong Bases)

Notes:

Problems:

Tips: 1) identify nucleophile; 2) draw product of Nucleophile addition; 3) see whether there are any good leaving groups; if yes: best LG leaves. if no: will have to protonate O to make a neutral compound.

(A) RCOCl $\xrightarrow{CH_3NH_2}$ RC(O)N(H)CH₃ + Cl⊖

Base picks up H^+

(B) RCOOR $\xrightarrow[HCl]{xs.\ CH_3OH}$

(C) RCOR $\xrightarrow{CH_3MgBr}$ R₂C(CH₃)O⁻ $\xrightarrow{H_2O}$ R₂C(CH₃)OH

(D) RCHO + ⊖≡—CH_3 ⟶ RCH(O⁻)C≡C—CH₃ $\xrightarrow{H_2O}$ RCH(OH)C≡C—CH_3

Notes:

* Base + Alkyne = Good Nu⁻

* R⊖, H⊖, NEVER LEAVE
OR⊖, OH⊖, $NH_2^{\ominus}$

Problems (cont'd):

(A) $RC(=O)OC(=O)R \xrightarrow{H_2O}$

(B) $RC(=O)OH \xrightarrow{CH_3NH_2}$ $RC(=O)NHCH_3 + H_2O$

(C) $RC(=O)OH \xrightarrow{HCl}$

Notes:

Observations

From the problems on the preceding slides, we learn

Observation 1: Only weak bases are good leaving groups; thus:

(A)

Observation 2: Acids and bases might be needed to move protons around to make good leaving groups or to facilitate Nu attack:

(B)

Observation 3: Carboxylic acids require special attention because they are acidic themselves:

(C)

Notes:

Relative Reactivities

Because each of the various carbonyl-containing functional groups has a different substituent, their reactivities to the initial nucleophilic addition vary as well:

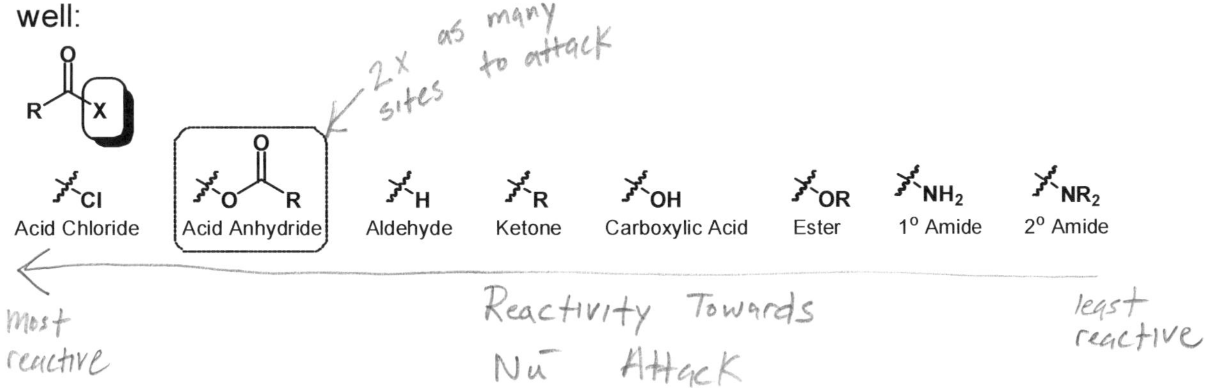

(A) Stabilize cation in E.A.S. Rxn, however, here, avoid stabilizing the $C^{\delta+}$

Notes:

Relative Reactivities

Another notable point is that some FGs have other contributing resonance structures:

A

B

Resonance contributors *Resonance hybrid*

Esters will have similar resonance structures wherein an O lone pair moves

Acid (or metal coordination) can increase the reactivity of a carbonyl:

C

much more reactive b/c of ⊕ charge

Notes:

Ester Hydrolysis I: Acid-catalyzed

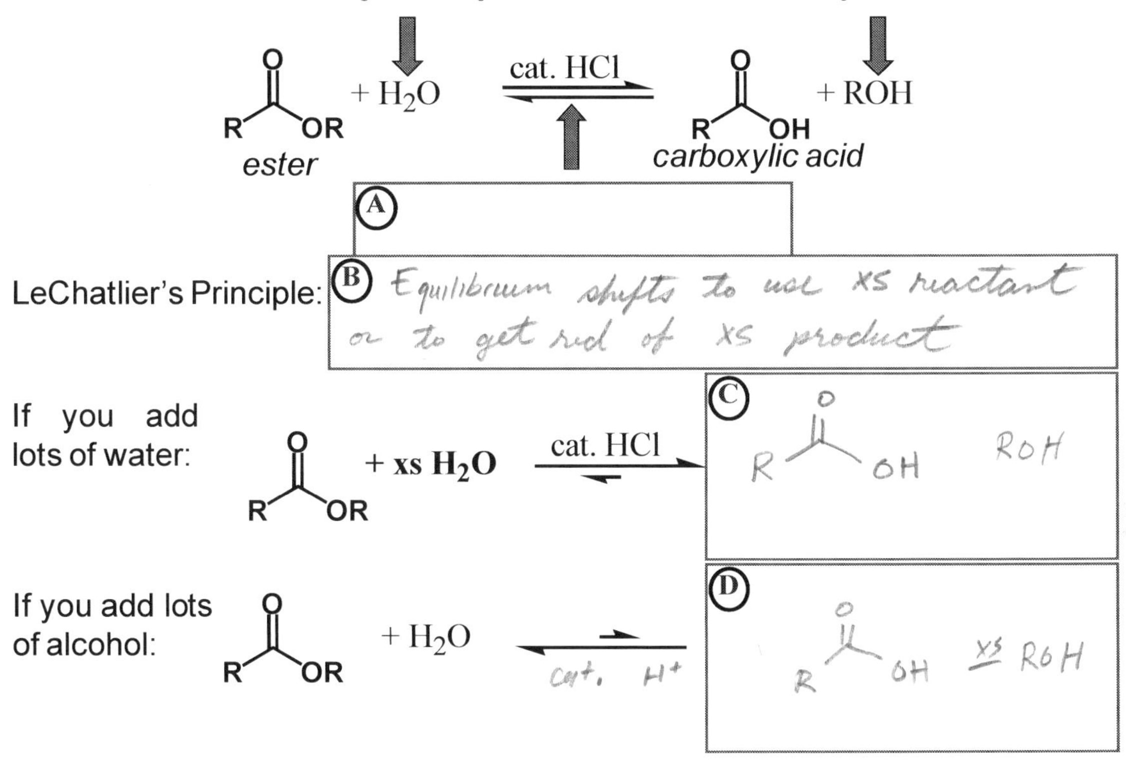

O
R OR
+ H₂O
cat. HCl
O
R OH
+ ROH
ester
carboxylic acid
A
LeChatlier's Principle:
B
Equilibrium shifts to use xs reactant or to get rid of xs product
If you add lots of water:
+ xs H₂O
cat. HCl
C
R OH
ROH
If you add lots of alcohol:
+ H₂O
cat. H+
D
R OH
xs ROH

Notes:

Acid-catalyzed ester hydrolysis: mechanism

Consider the mechanism for acid-catalyzed ester hydrolysis:

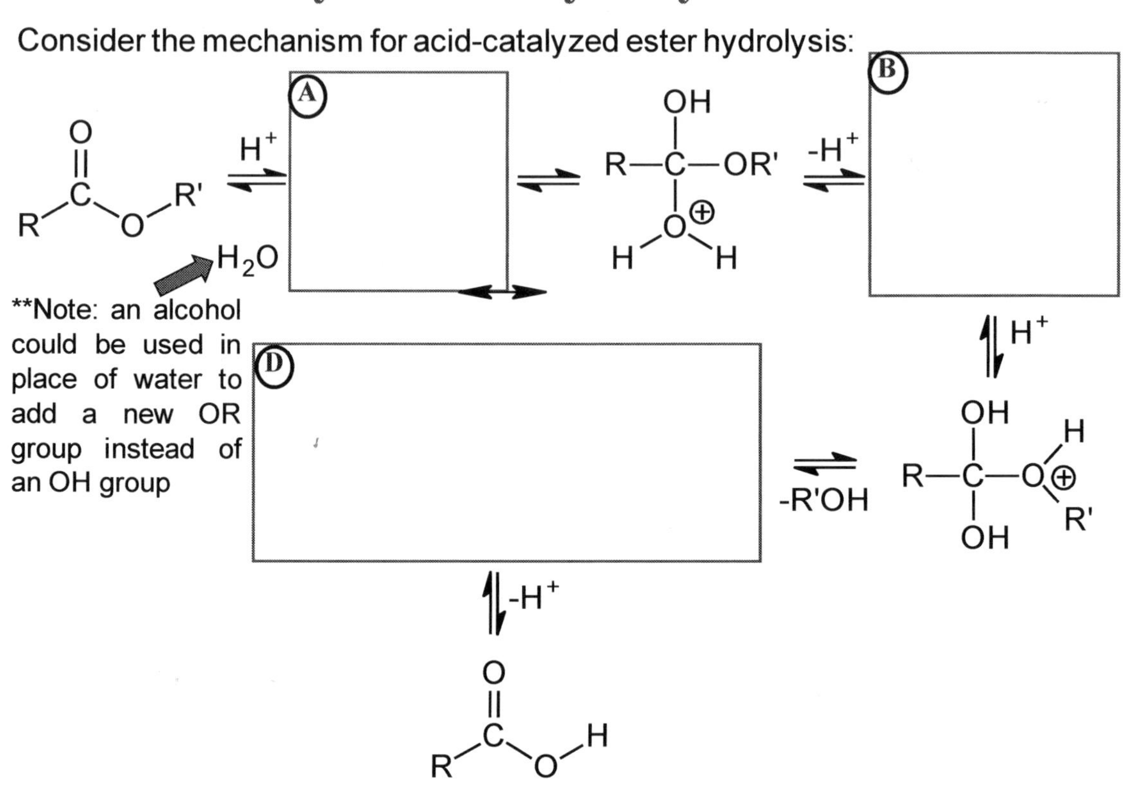

**Note: an alcohol could be used in place of water to add a new OR group instead of an OH group

Notes:

Base-promoted ester hydrolysis

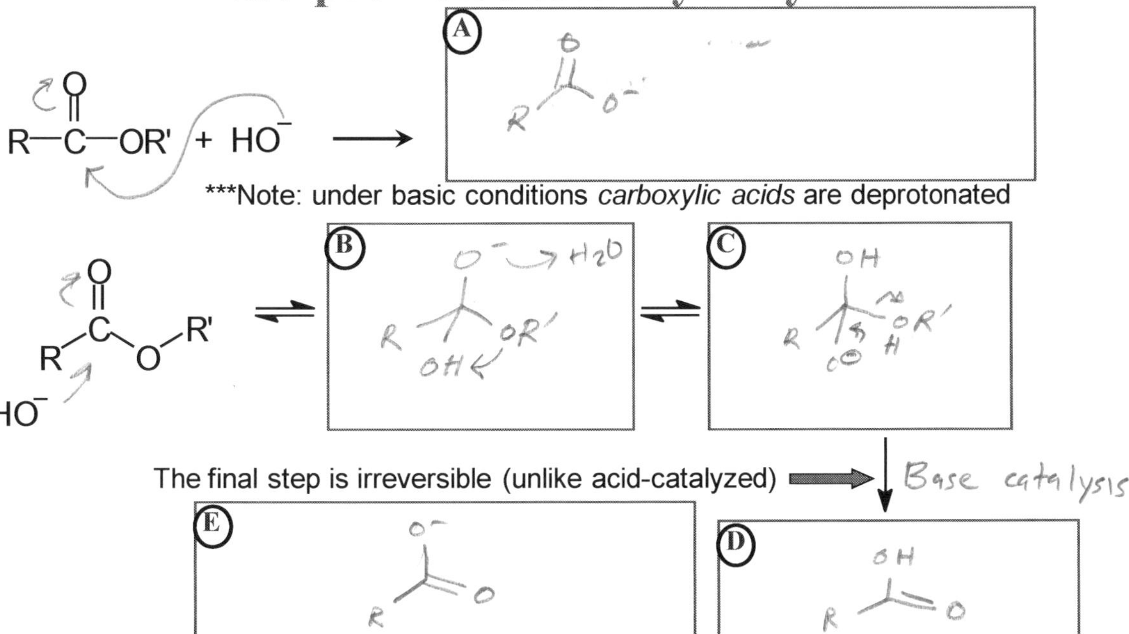
A
R—C(=O)—OR' + HO⁻
***Note: under basic conditions *carboxylic acids* are deprotonated
B
C
The final step is irreversible (unlike acid-catalyzed)
Base catalysis
E
D
XS HOR'

Notes:

Ammonolysis: the base doesn't have to be OH^-

If we replace the hydroxide with another nucleophilic base such as an amine or ammonia, we can make amides from esters:

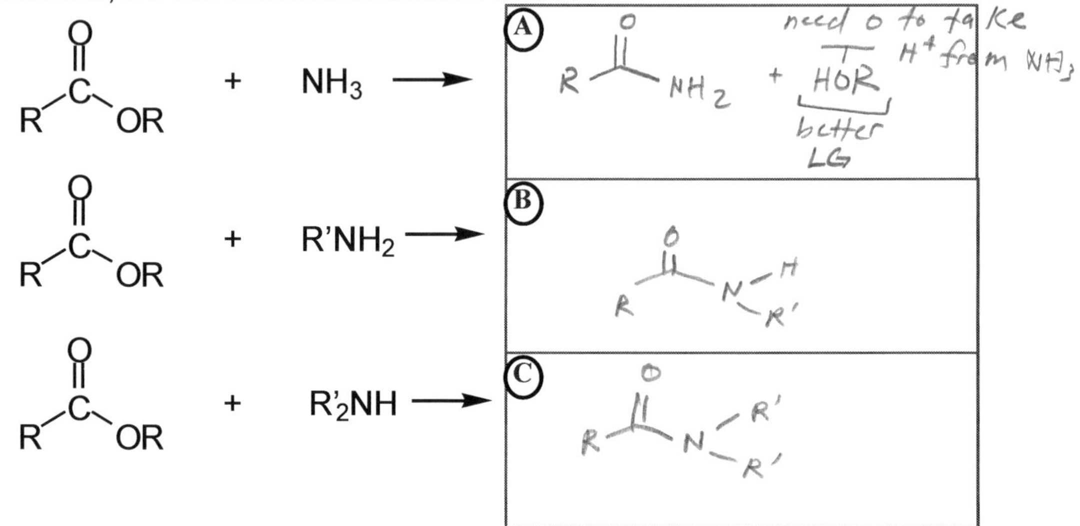

Note: these reactions are reversible (the amides are *not* deprotonated), so:

C CANNOT USE NR_3 (N needs H)

AND USUALLY you can distill ROH out during Rxn

Notes:

Amide hydrolysis

In fact, amides react with alcohols or water similarly to the way we saw for esters:

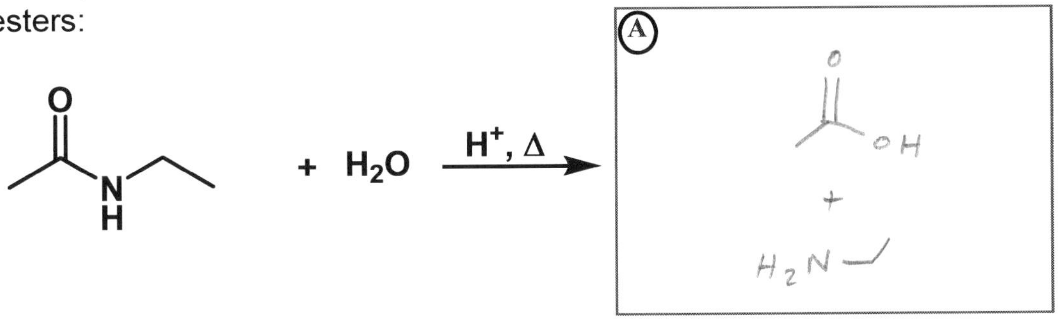

+ H_2O $\xrightarrow{H^+, \Delta}$ (A) + H_2N

once made CANNOT REACT

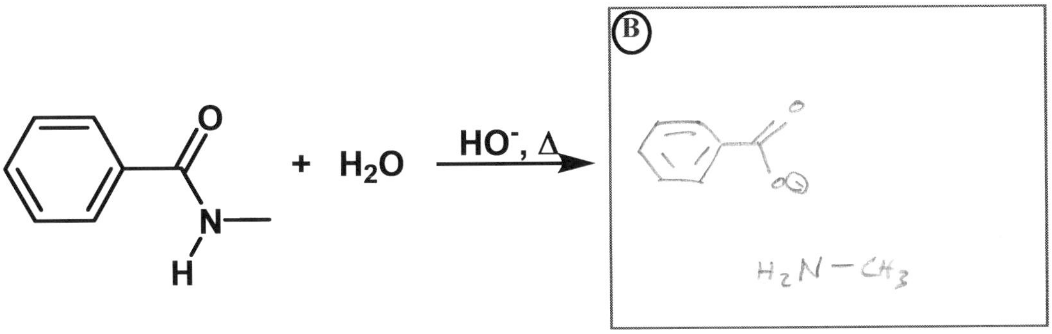

+ H_2O $\xrightarrow{HO^-, \Delta}$ (B) H_2N-CH_3

Notes:

Acid-catalyzed amide hydrolysis

Amide acid-catalyzed hydrolysis employs the same mechanism as for esters:

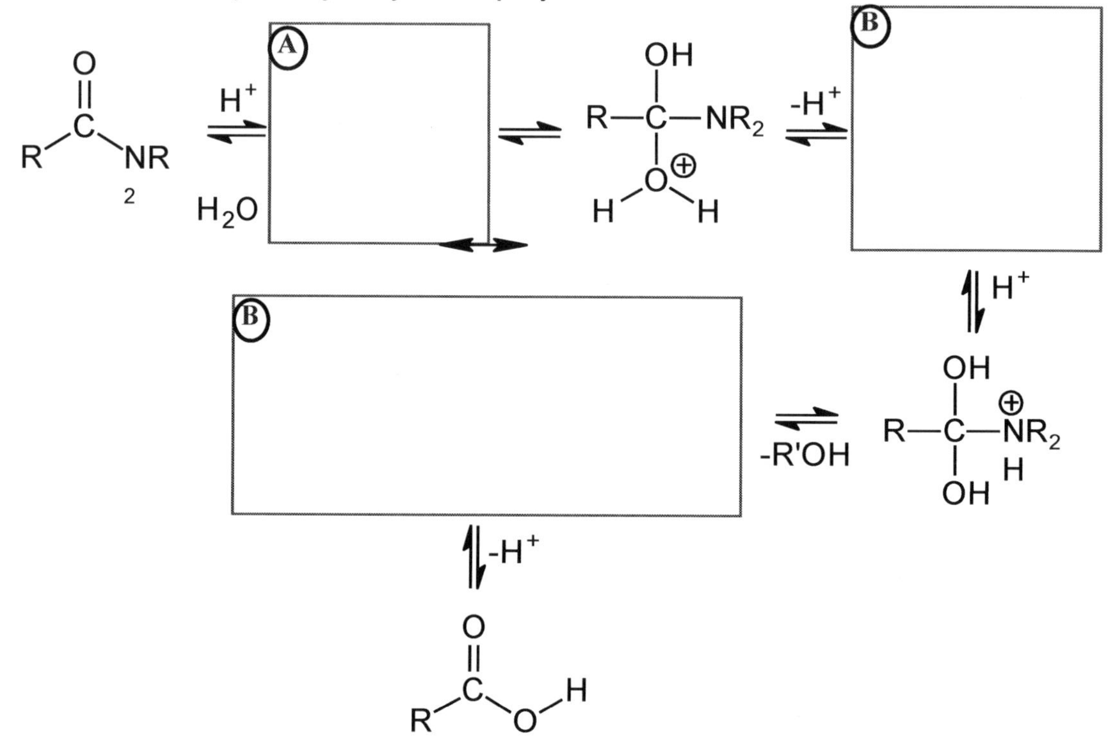

Notes:

Use of amide hydrolysis: Gabriel Synthesis

Preparation of primary amines from alkyl bromides is very useful; however, if one uses ammonia as the nucleophile, up to four R groups can add to N:

The Gabriel Synthesis yields 1° amines by protecting two sites of the N for the S_N2 reaction, then deprotection by acid hydrolysis:

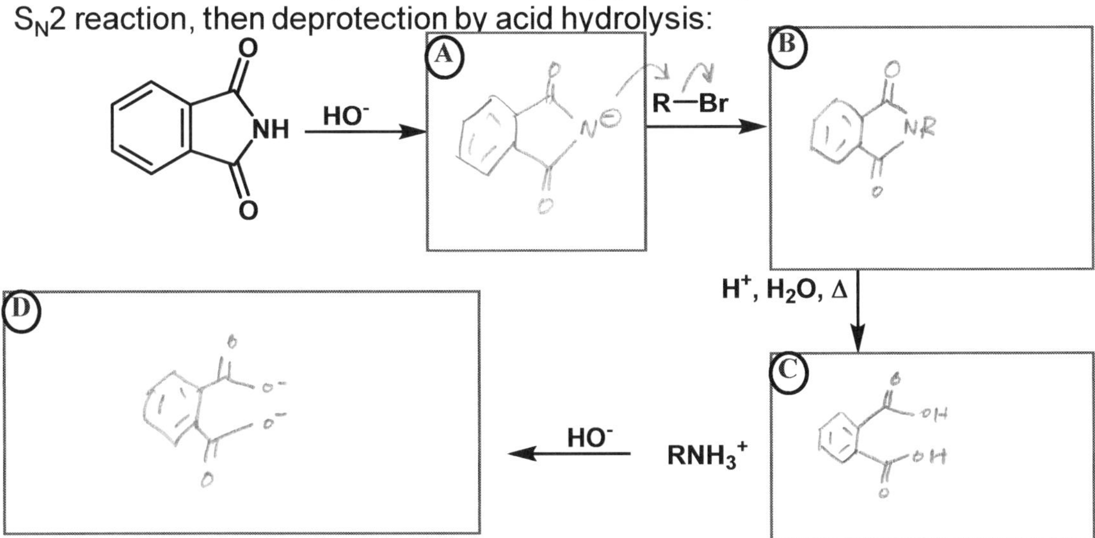

Notes:

① NaOH ← deprotonate
② CH_3Br ← S_N2
③ H^+, H_2O, Δ ← Amide hydrolysis
④ NaOH ← Workup step

Some rxns don't work as well with carboxylic acids

Some of the reactions that wok well on amides or esters may not work as well on carboxylic acids because of their acidity:

(A) $CH_3COOH + NH_2R \rightarrow CH_3COO^- + {}^{\oplus}NH_3R \xrightarrow[225°C]{\Delta}$ CH_3CONHR

thermal Decomposition

Thus, when a carboxylic acid reacts with a nucleophile, ALWAYS consider:

(B) ONLY Strong $Nu^{\ominus}$ can attack the Carboxylate (CH_3COO^-)

Notes:

What about aldehydes and ketones?

Aldehydes and ketones can react with water or alcohols similarly to what we've seen for esters and amides, but if we remove the water as it is made, we can stop the reaction at the **ketal** (for ketones) or **acetal** (for aldehydes):

O
H^+ H^+
^+OH
CH_3OH
A
OH
CH_3OH ⊕
H^+ H^+
B
OH
OR
H^+ H^+
E
OR
OCH_3
Ketal
H^+ H^+
D
H ⊕ O R
OCH_3
CH_3OH
$+OCH_3$
C
H_2O ⊕
OR

Notes:

O + 2 HOR → OR / OR Ketal + H_2O ← taken out as rxn proceeds

Carboxylic acids: the bad LG problem

Acid chlorides are good starting materials because they can undergo nucleophilic acyl substitution easily (the Cl is a good leaving group). We've seen that esters and amides can undergo net substitution reactions as well; however, carboxylic acids are easily deprotonated and the –OH group is not a ready leaving group like Cl. For this reason, chemists have developed other ways to **activate** the carboxylic acid. This is very similar to how we activated the OH to be a good leaving group when we studied reactions of alcohols; in fact some of the same reagents are used:

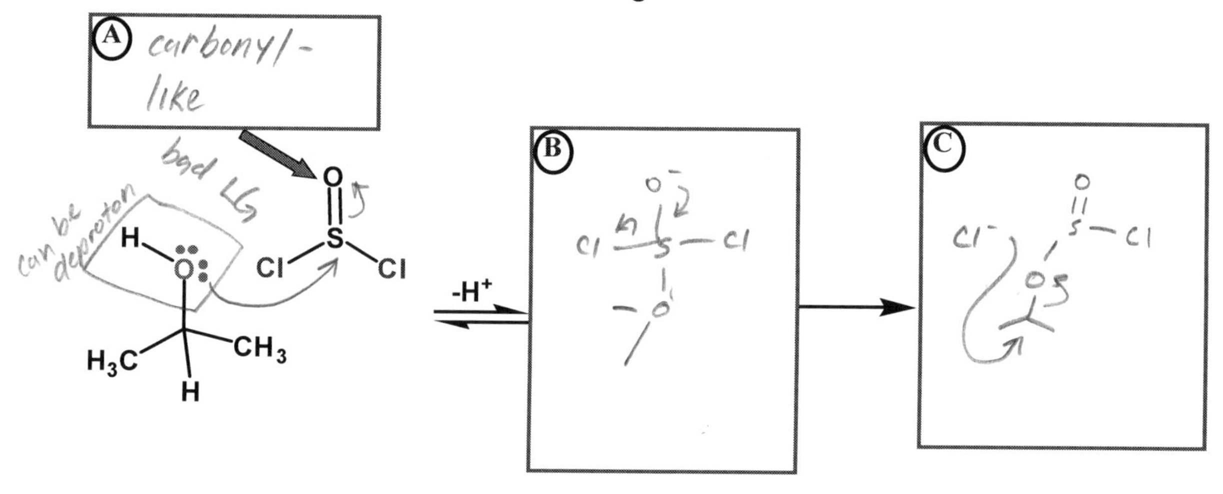

Notes:

Carboxylic acids: the bad LG problem

Using PX_3 (X = Cl or Br):

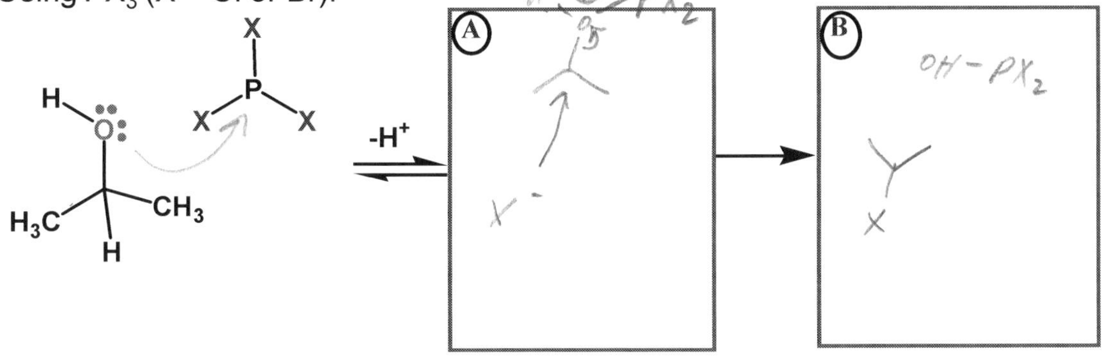

H, O, X, P, X, X, H_3C, CH_3, H

$-H^+$

A

B

So let's look at how carboxylic acids react with thionyl chloride or PCl_3:

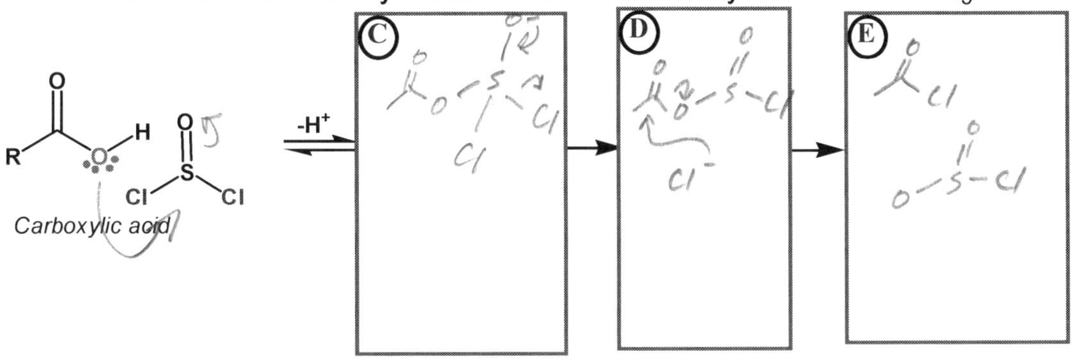

O, R, O, H, O, S, Cl, Cl

Carboxylic acid

$-H^+$

C

D

E

Notes:

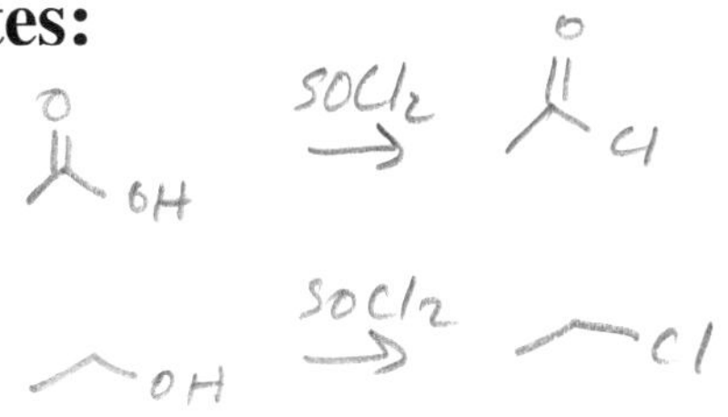

Carboxylic acids: the bad LG problem

Using PX_3 (X = Cl or Br):

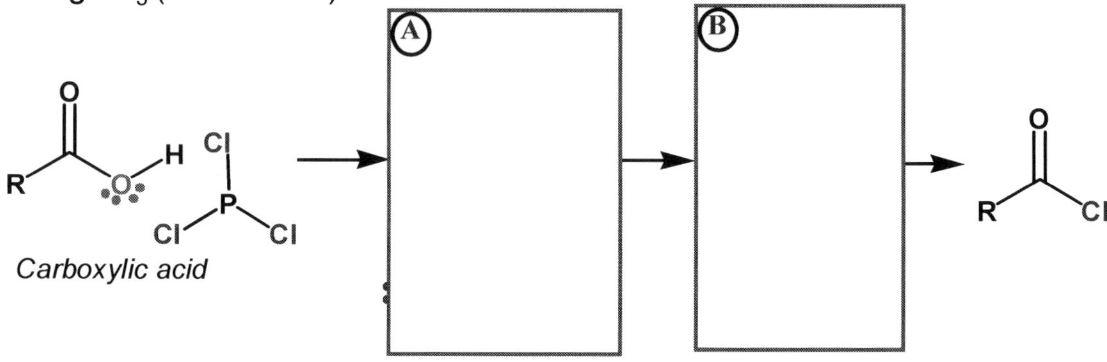

So the purpose of either $SOCl_2$ or PCl_2 is to:

Ⓒ

The net reaction is:

Ⓒ

Notes:

Specific for 1 amides: Prep. of a nitrile

One more reaction in your text that a 1° amide may undergo is dehydration to form a nitrile:

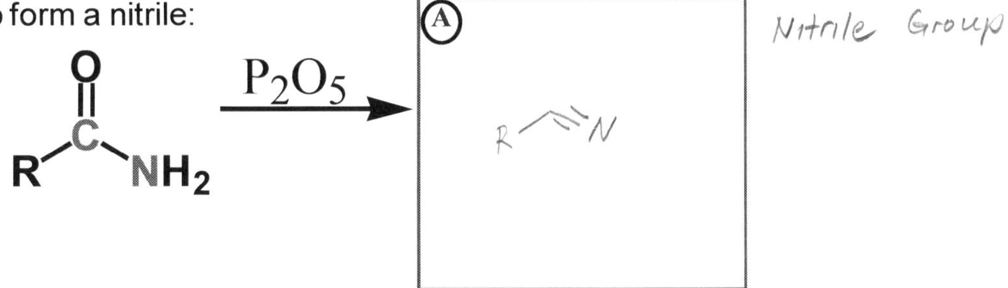

We will not go into the detailed mechanism, but it is worth noting that the nitrile can itself be converted to a carboxylic acid upon hydrolysis:

Notes:

* P_2O_5 very good dehydrating agent

Reactions of Nitriles

You will notice that the polar C≡N bond pi bond is susceptible to nucleophilic attack in a fashion similar to that observed for ester hydrolysis:

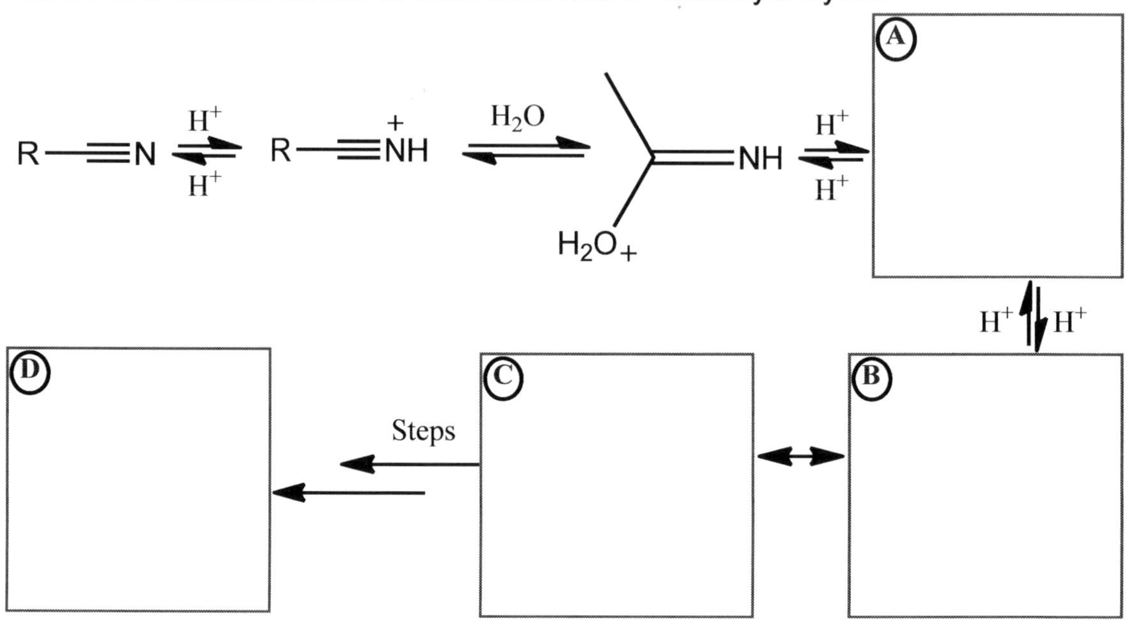

Notes:

Recap: Reactions

Reaction 1: Acid anhydrides and acid chlorides

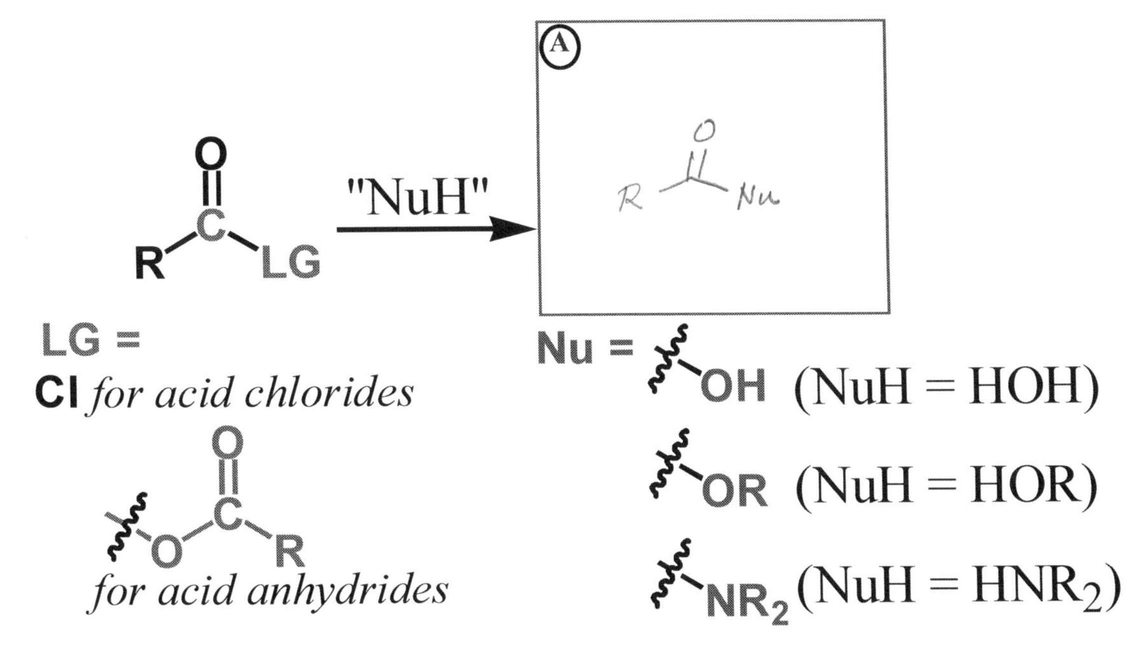

Notes:

Recap: Reactions

Reaction 2: Ester/carboxylic acid equilibrium in presence of acid catalyst

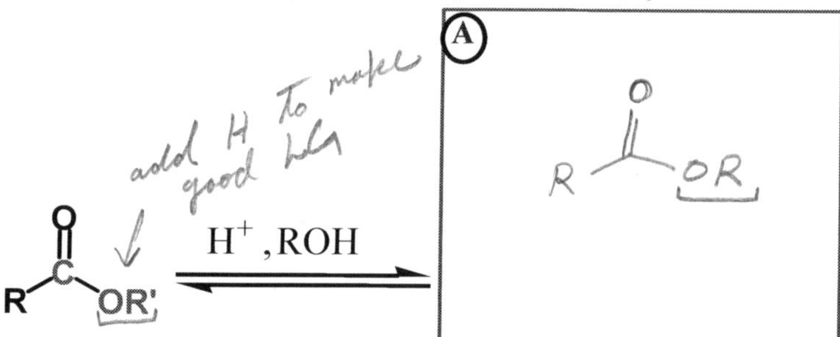

R' = H: a *carboxylic acid* is the starting material
R = hydrocarbon: an *ester* is the starting material

R = H (water!): makes a *carboxylic acid*
R = a different R group than the one that is already on the ester: makes an *ester*

*****Add excess ROH to push equilibrium to the right*****

Notes:

Recap: Reactions

Reaction 3: Ester reactions in presence of a base

3a: The base is hydroxide (*base catalyzed hydrolysis):

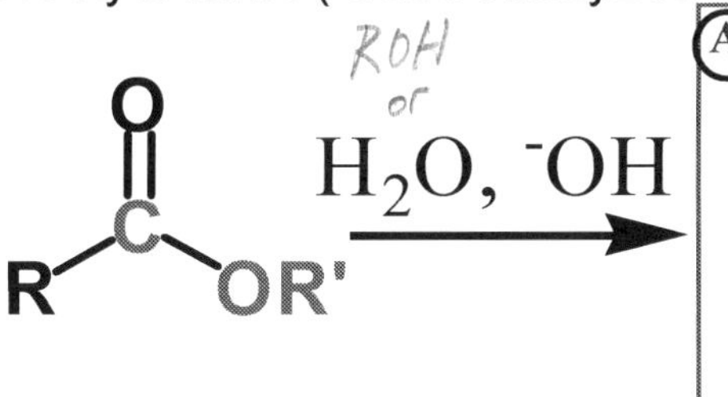

A

O
R
OR

3b: The base is an amine (ammonolysis):

O
C
R''
OR'
HNR_2

(R can be H or a hydrocarbon)

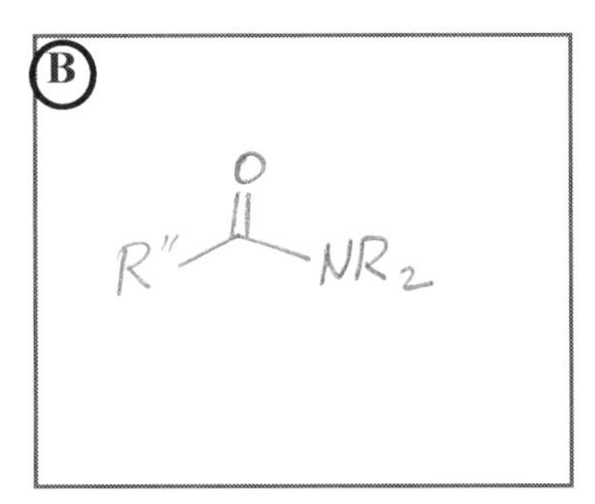

Notes:

Recap: Reactions

Reaction 4: Aldehyde or Ketone with

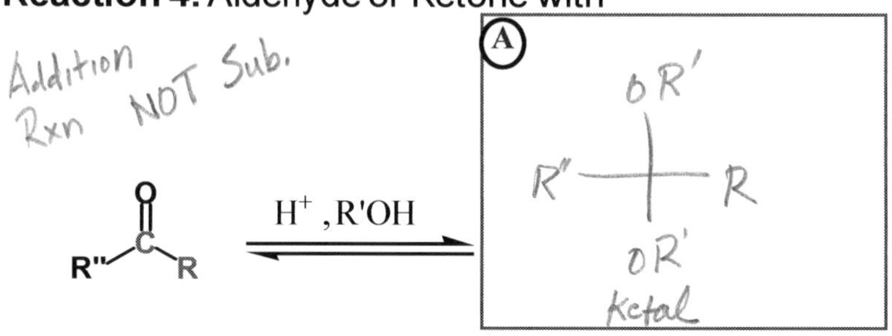

*****Important to remove water as it forms to push reaction to the right*****

R' = H: a *carboxylic acid* is the starting material
R = hydrocarbon: an *ester* is the starting material

R' = H: an *acetal* is the product
R = hydrocarbon: a *ketal* is the product

Reaction 5: Carboxylic acid activation (make an acid chloride)

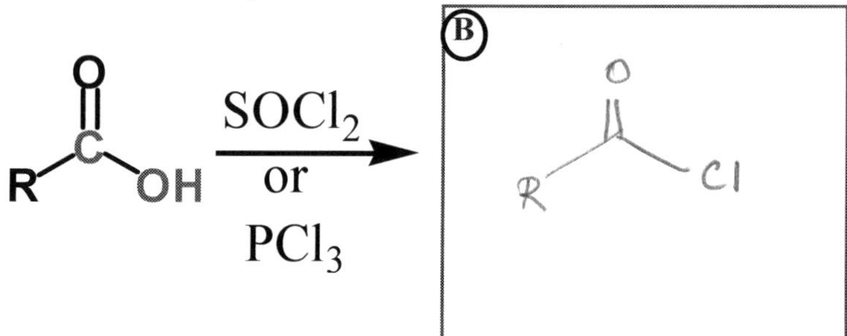

Notes:

* if no LG is good then can only add groups to starting material

Recap: Reactions

✳**Reaction 6:** Dehydration of an amide to make a nitrile:

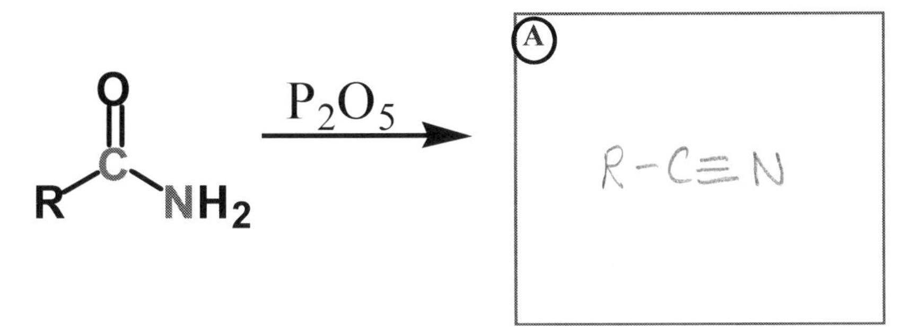

✳**Reaction 7:** Hydrolysis of a nitrile to make a carboxylic acid:

R—C≡N $\xrightarrow[\Delta]{H+,\ H_2O}$ (B)

Notes:

Lecture Set 7:
Reactions of Aldehydes, Ketones and Carboxylic Acid Derivatives

Suggested Reading:

Suggested Problems:

Nomenclature of Aldehydes

Aldehyde nomenclature is simple because an aldehyde can only be at the **end** of a chain; so name the parent chain and end the name in "-al":

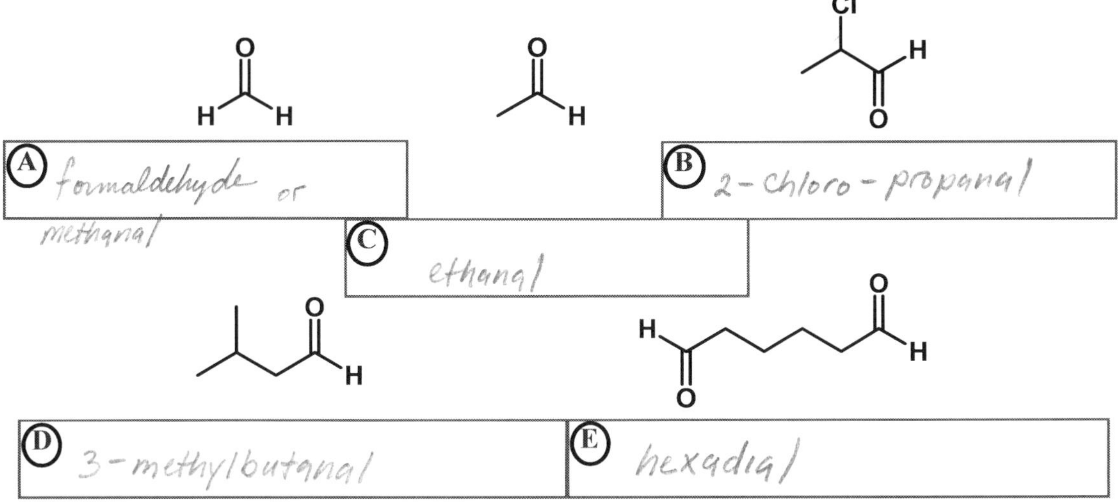

An aldehyde coming off of a ring is named as a substituent "carbaldehyde":

F Trans-2-methylcyclohexane carbaldehyde

carbaldehyde

Notes:

Nomenclature of Ketones and Aldehydes

The carbonyl group of a ketone can be in the middle of the molecule, so put a number in from to the parent chain name to tell which C has the O and put an "-one" suffix on (similar to rule for alcohol, except alcohol has an "-ol" ending):

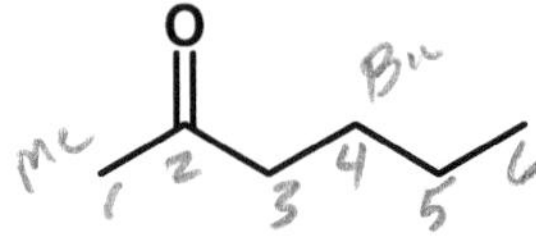

2-hexanone

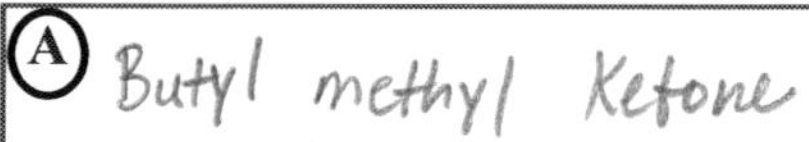

(A) Butyl methyl Ketone	(B) 2-methyl-3-heptanone

The second name listed is the "common name" of the ketone. You can make a common name by listing the names for the chains on each side of the carbonyl unit (alphabetically) followed by "ketone" (quite similar to the way we learned to make common names for ethers).

Priorities:

(C) Aldehyde > Ketone > alcohol

Notes:

Some General Trends in Carbonyl Reactivity

Recall one of the two general reactions for carbonyls:

II.

X–C(=O)–Y + $Nu^{\ominus}$ ⟶ (A) X–C($O^{\ominus}$)(Nu)–Y *No Good LG* —H^+ or H_2O⟶ (B) Y–C(OH)(Nu)–X

****This is the key reaction type to be examined in this chapter for ketones and aldehydes**

Now that we know that the general reaction of interest is, all we have to do is look at different nucleophiles that will work to make useful products ...

Notes:

Organometallics and Metal Hydrides

A C–M or H–M (M = a metal) bond is very polar and places substantial negative charge on the C or H that is directly bonded to the M. We learned about organometallic compounds where the C was bound to Mg (Grignard reagents), Li, Cu (Gilman reagents), B (Suzuki Reaction), and Sn (Stille reaction). In the current context, the important points are:

1. For the purpose of mechanisms:

(A) R-MgX Al-H hydrides } Al⊕ ⊖H } unstable anions powerful Bases Super Nu⊖

R-Li

R⊖ ⊕MgX or Li⊕

2. Consider the nucleophile strength:

(B)

Notes:

Organometallics and Hydrides

In this chapter, we will explore the reactivity of a few organometallic species.

1. Reagents that give nucleophiles with minus charge on C

Grignard Reagents: *Acetylide anions:* *Cyanide anions:*

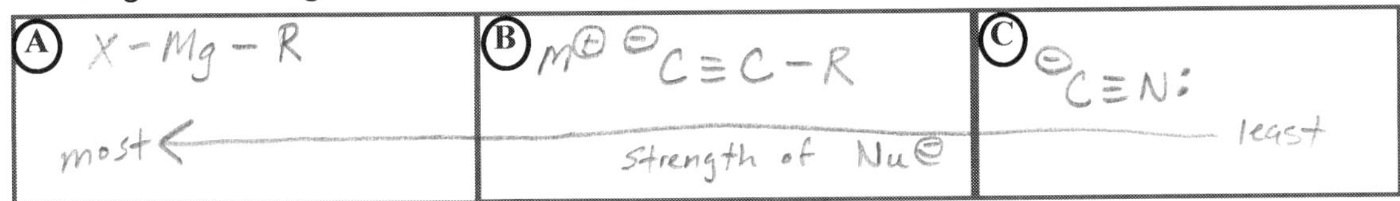

2. Reagents that give H⁻ as nucleophiles (H⁻ is called ***Hydride***)

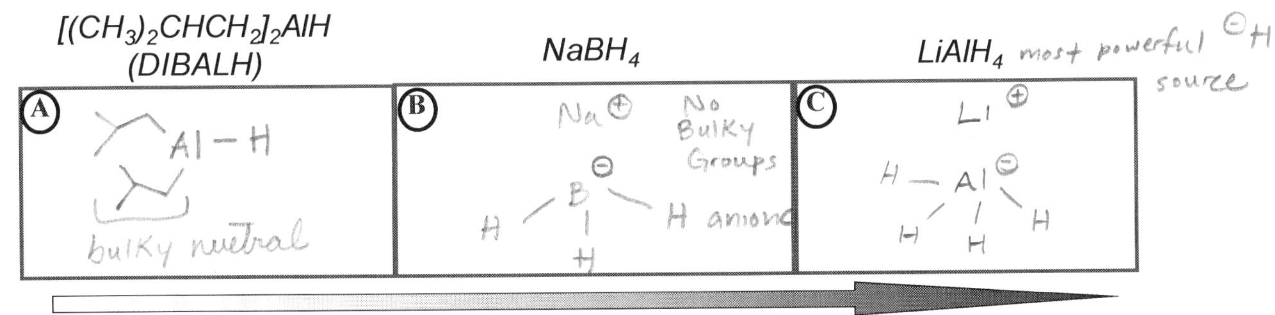

You should be able to predict the reactivity trend on the basis of charge, H-M bond polarity and steric bulk

Notes:

Reactions of Ketone or Aldehyde

Whether the nucleophile is R⁻ or H⁻, both aldehydes and ketones react via the general mechanism:

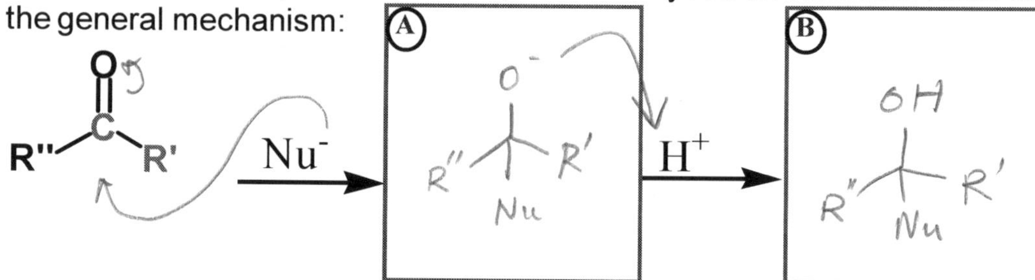

R' can be a hydrocarbon (starting material is a ketone) or a hydrogen (starting material is an aldehyde)

Nu can be a Grignard, organolithium, acetylide, cyanide, or hydride source

Simple recognition of this general reaction allows you to successfully predict the major products for a number of reactions that are presented in your text...

Notes:

Problem

Example: Give the major product for each of the following reactions.

(A)

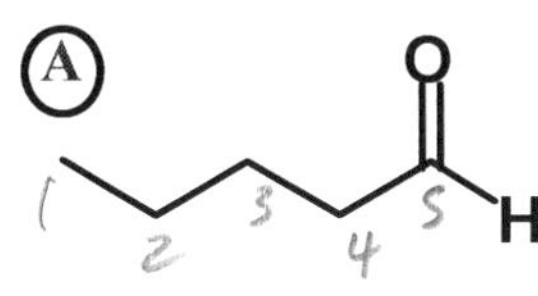

1. $NaBH_4$
2. H^+

$H^{\ominus}$ = Nu

(B)

1. CH_3CH_2MgBr
2. H^+

= Nu

(C)

1. $LiAlH_4$
2. H^+

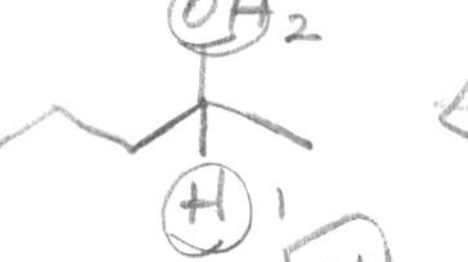

← Doesn't matter if aldehyde or ketone

(D)

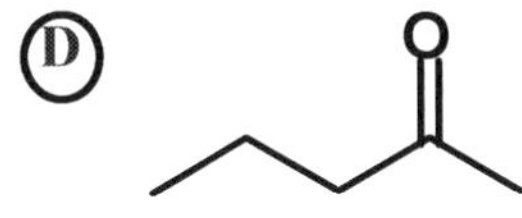

1. $\ominus \equiv$—Ph
2. H^+

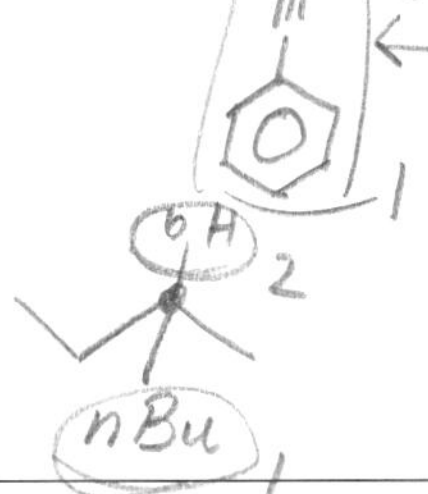

← can react alcohol

← or can react via triple bond

(E)

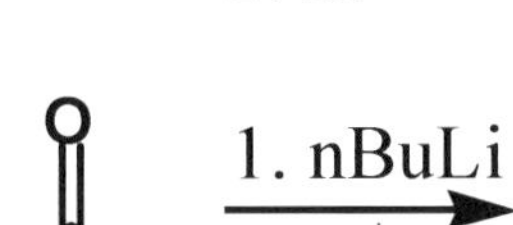

1. nBuLi
2. H^+

Notes:

(A)

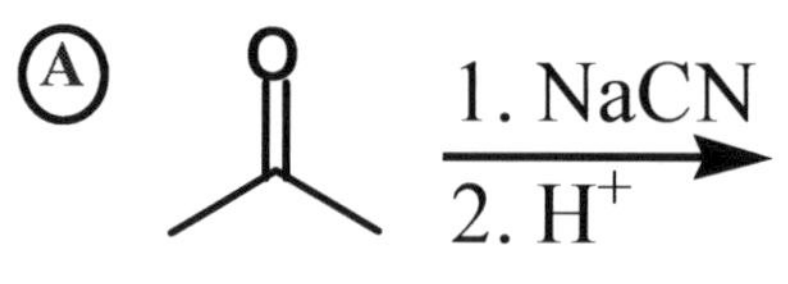

A cyanohydrin

(B)

1. PhMgI

2. H^+

(C)

H

1. CH_3MgCl

2. H^+

(D)

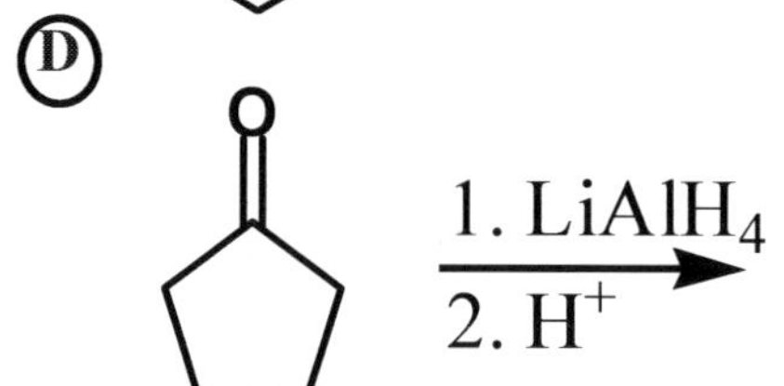

(E)

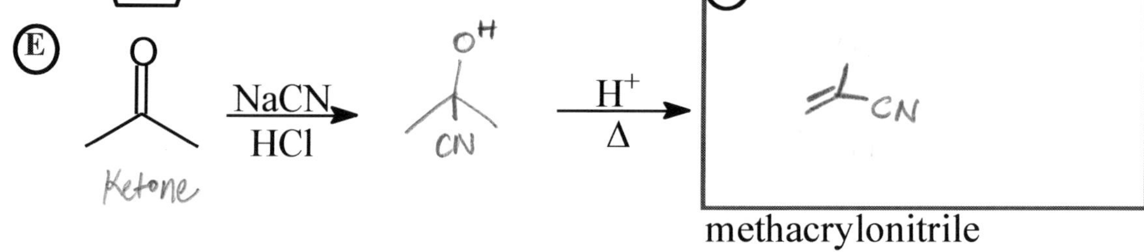

methacrylonitrile

Notes:

What about all the other functional groups??

Because R⁻ or H⁻ are powerful nucleophiles, they will react with other carbonyl compounds, not just ketones and aldehydes. We need to carefully assess what the products of such reactions will be. Consider the mechanism:

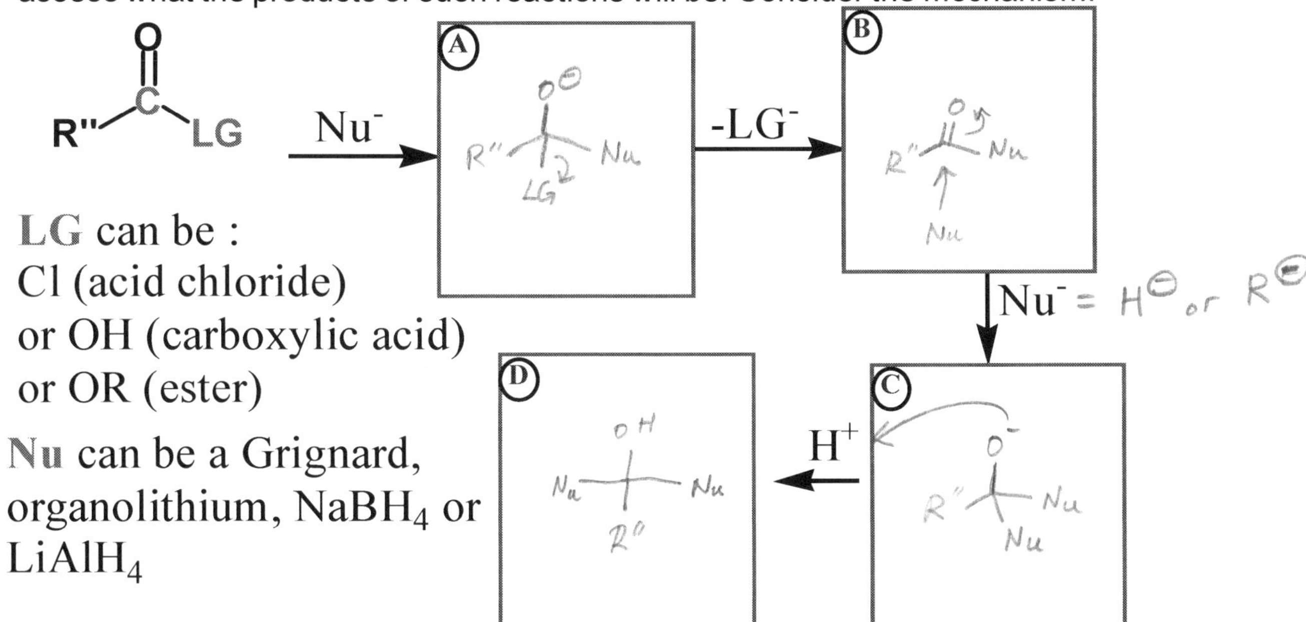

LG can be :
Cl (acid chloride)
or OH (carboxylic acid)
or OR (ester)

Nu can be a Grignard, organolithium, $NaBH_4$ or $LiAlH_4$

Notes:

R or H can add twice to some carbonyls

We are usually interested in the net reaction (major products). In this case, note that for acid chlorides, esters, and carboxylic acids, R^- or H^- add twice:

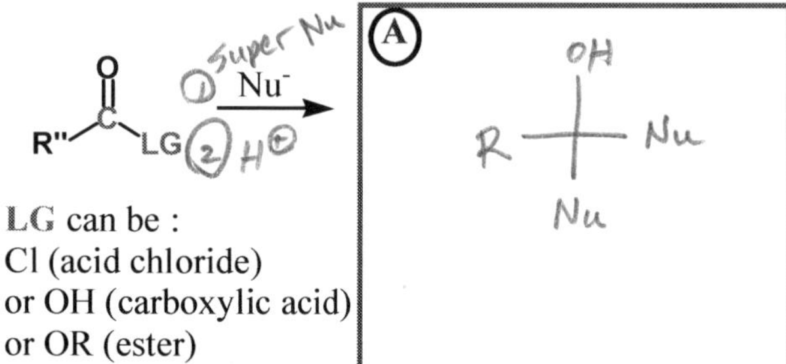

LG can be :
Cl (acid chloride)
or OH (carboxylic acid)
or OR (ester)
Nu can be a Grignard, organolithium, $NaBH_4$ or $LiAlH_4$

One important note for carboxylic acids:

(B)

Notes:

DIBALH is special ...

Recall when we examined the hydride sources to be used in this chapter, we noted that DIBALH is the least reactive of the hydride sources under discussion. In fact, if we use DIBALH as the hydride source and keep the temperature low (typically -78 C, we are able to stop the reduction of an ester at the aldehyde:

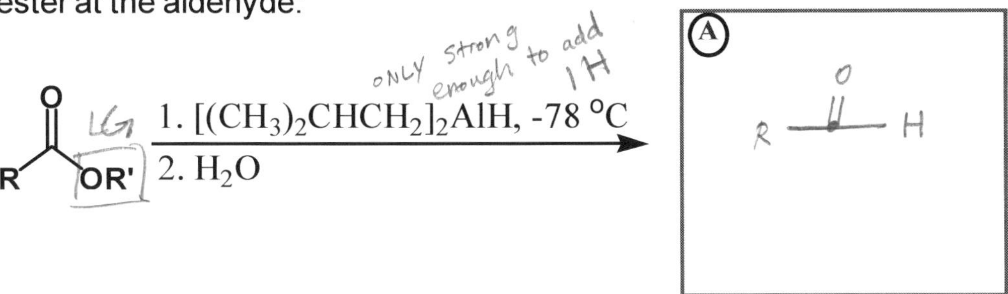

This is an important reaction for preparing aldehydes and is worth committing to memory.

Notes:

Why are we ignoring amides ???

The situation is slightly different when an amine is the reactant, but if we carefully follow the mechanism and examine the best leaving group in each case, we will arrive at the correct product:

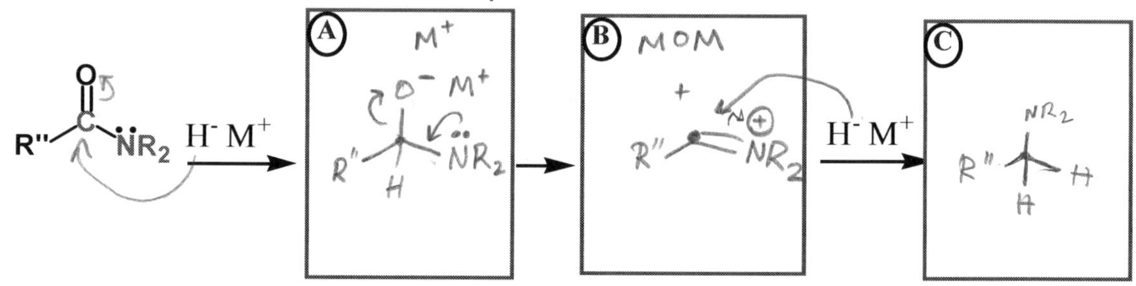

So the net reaction is the preparation of an **amine** (NOT an am*ide*):

$R''C(=O)\ddot{N}R_2 \xrightarrow{LiAlH_4}$ **D** R''—CH_2—NR_2

The transformation we observe from A to B is a key step for a couple other reactions presented in this chapter ...

Notes:

Reaction of ketones/aldehydes with 1 amines

When we react a ketone with a 1 amine, we form a species similar to the one in box A on the previous slide.

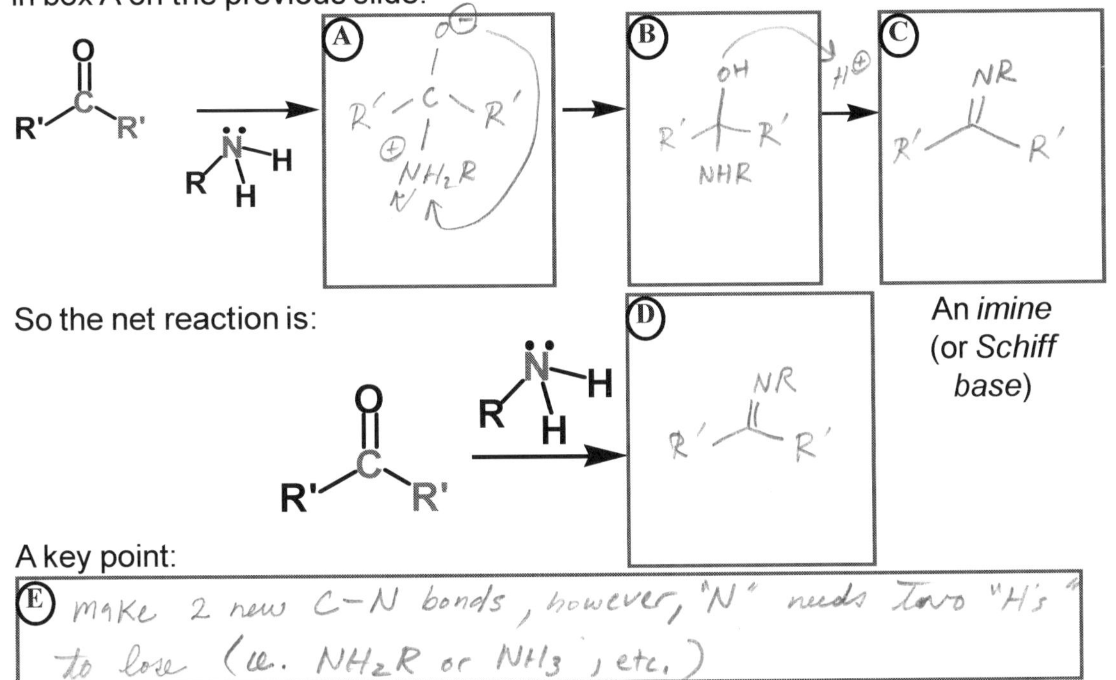

A key point:

E make 2 new C–N bonds, however, "N" needs two "H's" to lose (i.e. NH_2R or NH_3, etc.)

Notes:

Reaction of ketones/aldehydes with 2 amines

Secondary amines lack the second proton that has to leave in order to form the neutral imine, so if we react a ketone or aldehyde with a secondary amine, the initial stages of the reaction are similar to those observed on the previous slide. However, when it comes time for the second proton to leave, a different proton needs to be used:

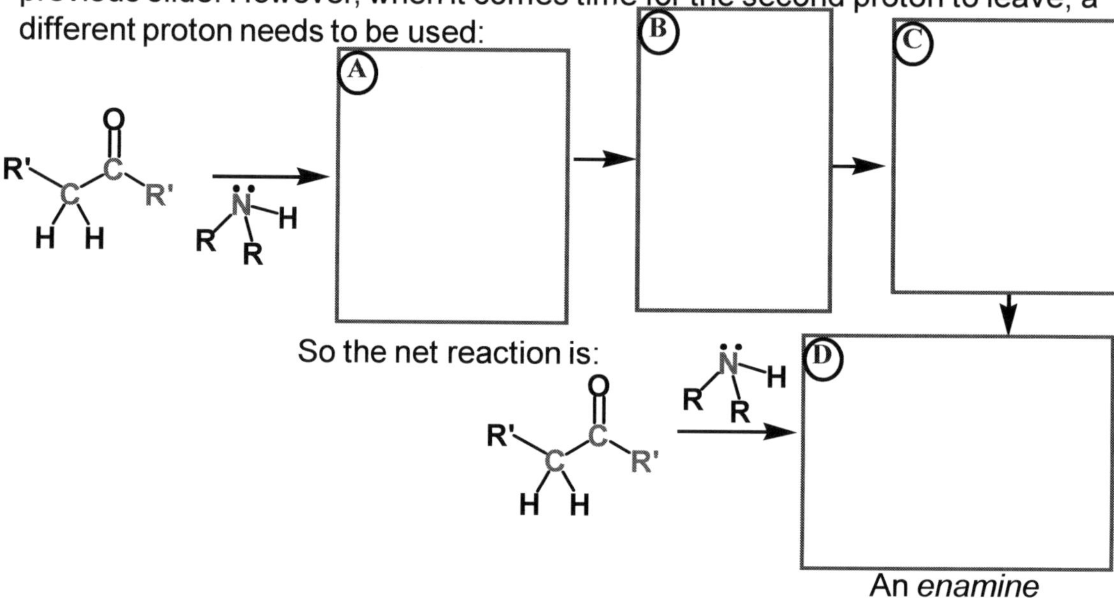

Notes:

Reaction of ketones/aldehydes with H_2NZ

Your text provides a few other reactions of ketones and aldehydes with other reagents of the form "H_2NZ" that are nominally similar to primary amines in their reactivity pattern:

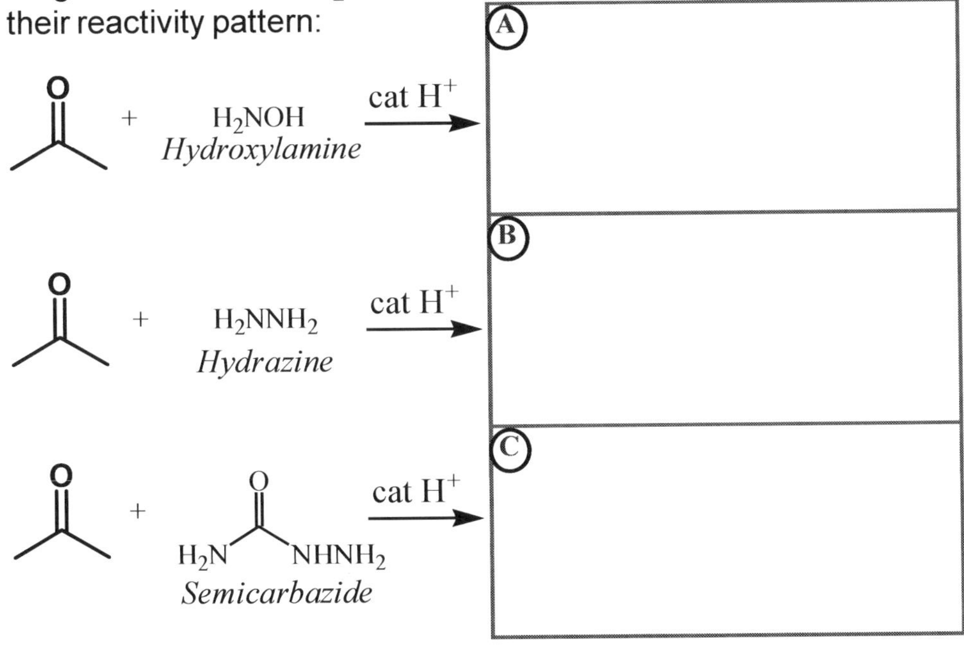

Notes:

Ketal/acetal formation

Recall the reaction we reviewed earlier:

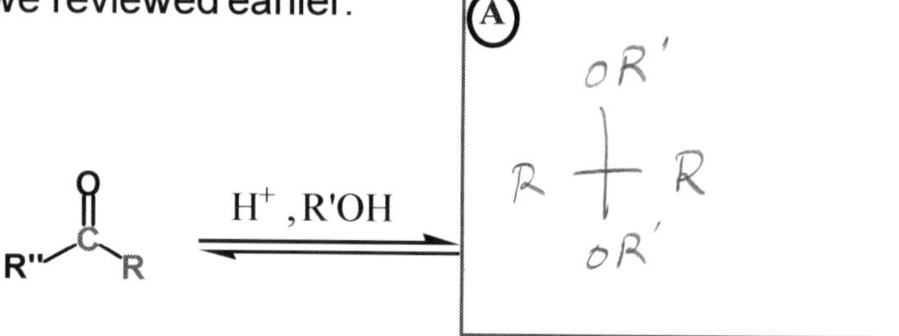

R' = H: a *carboxylic acid* is the starting material
R = hydrocarbon: an *ester* is the starting material

R' = H: an *acetal* is the product
R = hydrocarbon: a *ketal* is the product

The reaction is in equilibrium, so we can push it either way:

Ⓐ Remove H_2O to push forward.
Add " " " back.

Notes:

Ketal/acetal formation

(A) $\xrightarrow[\text{-}H_2O \text{ (distil.)}]{CH_3OH, \text{ cat. } H^+}$ OCH_3 OCH_3

(B) H O $\xrightarrow[\text{-}H_2O \text{ (distil.)}]{CH_3OH, \text{ cat. } H^+}$ H OCH_3 OCH_3

(C) O $\xrightarrow[\text{-}H_2O \text{ (distil.)}]{CH_3OH, \text{ cat. } H^+}$ OCH_3 OCH_3

(D) ☆ O HO OH, cat. H^+ $\xrightarrow[\text{-}H_2O \text{ (distil.)}]{}$ O O *

Notes:

Ketals/acetals as PROTECTING GROUPS

Acetals/ketals are treated in some detail by your text because of their importance as **protecting groups:**

Br

O

HO OH, cat. H^+

$-H_2O$
(distil.)

Susceptible to nucleophilic attack

A

Br

Protected from nucleophilic attack

⊖≡—R

B

R—≡

H^+, H_2O

C

R

O

Notes:

Combining Knowledge: the Wittig Reaction

If PPh_3 reacts with RBr, what is the product?

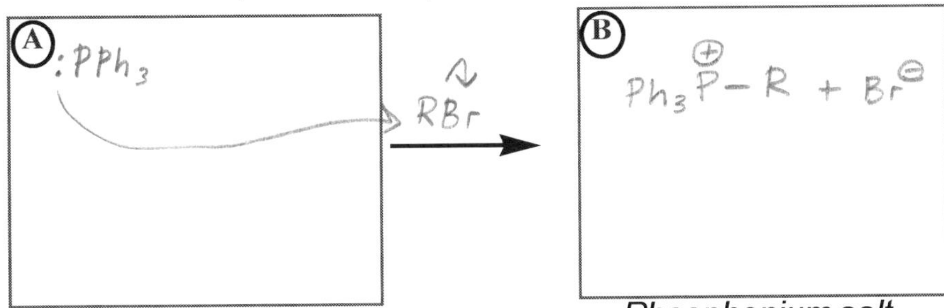

Phosphonium salt

If the product above is reacted with a strong base, what is the product (which proton is taken off by the base)?

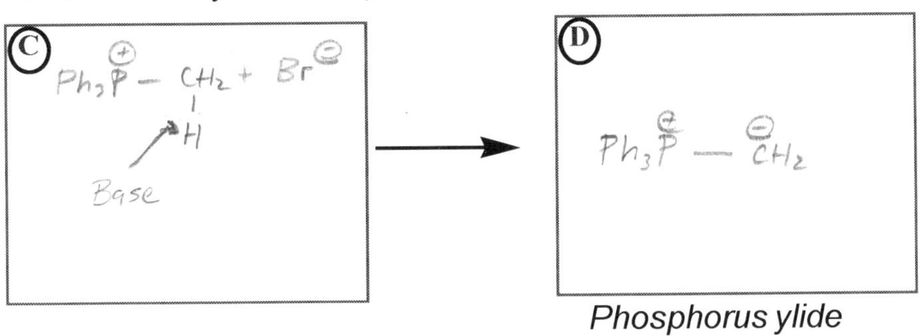

Phosphorus ylide

Notes:

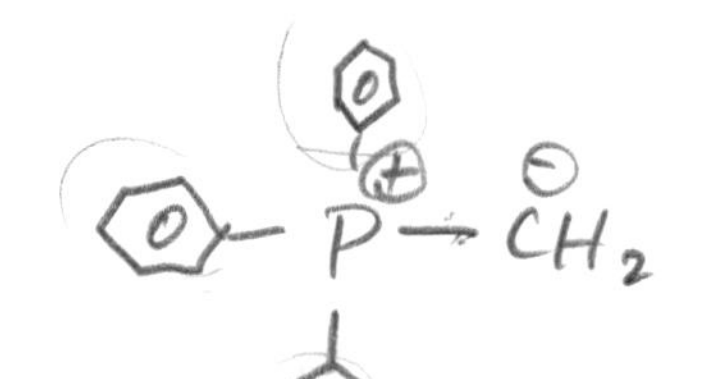

Combining Knowledge: the Wittig Reaction

This product can be drawn in a couple resonance structures:

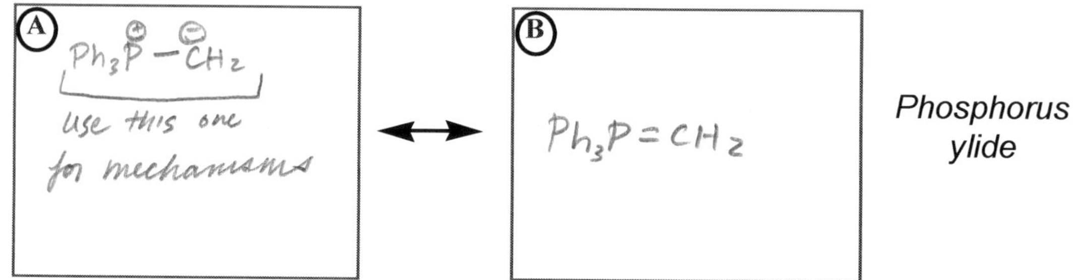

Phosphorus ylide

Now what happens if we use this compound as a nucleophile to react with an aldehyde or ketone?

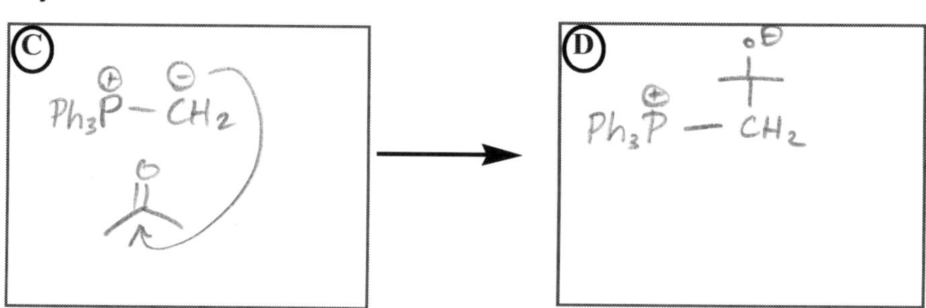

Notes:

Combining Knowledge: the Wittig Reaction

Now we have a molecule with both positive and negative charges close to one another; what happens?

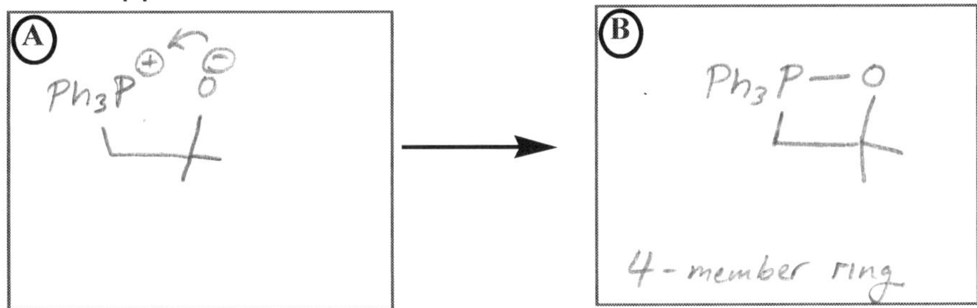

Now we have a four-membered ring; this suffers a lot of ring strain. How can we relieve ring strain?

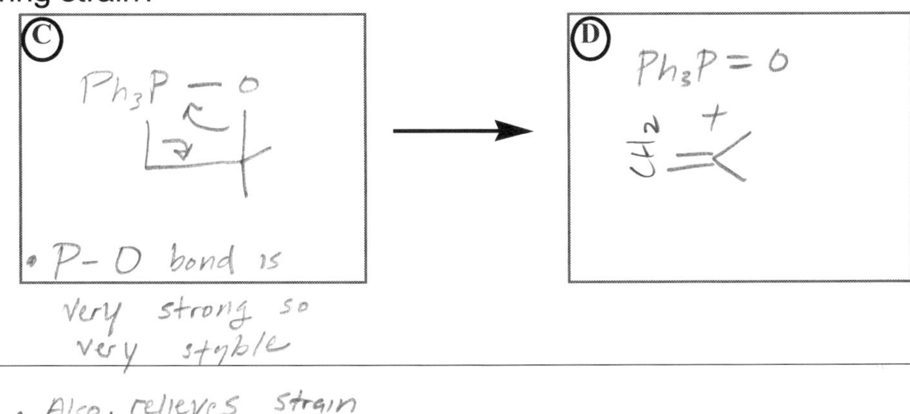

Notes:

- Also, relieves strain
- makes alkene

Combining Knowledge: the Wittig Reaction

Over the past few slides, we've worked through a series of steps required to carry out the **Wittig reaction**. Here is the net reaction as it is frequently shown in your text:

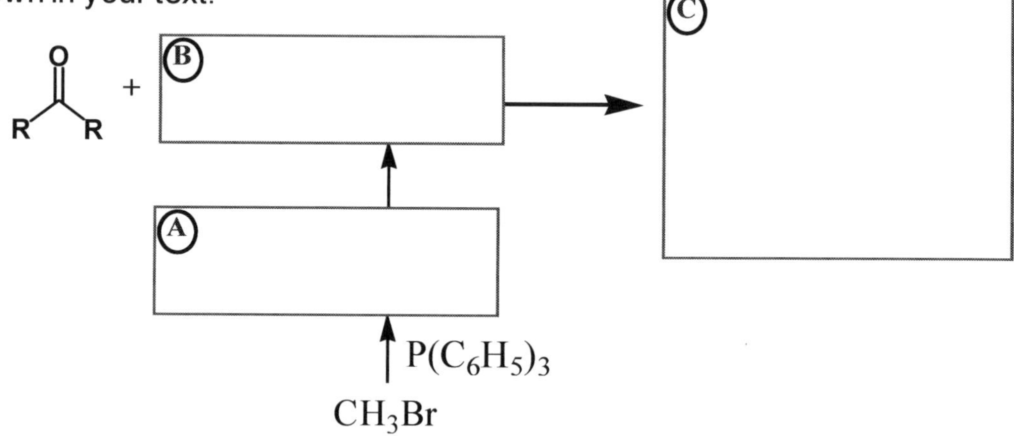

Note that although the preparation of the phosphorus ylide is often shown with the Wittig reaction, it is not part of the actual "Wittig reaction"; the term "Wittig reaction" refers to the reaction of the ylide with the carbonyl species.

Notes:

Lecture Set 8:
Reactions of Carbonyls at the α-Carbon

Suggested Reading:

Suggested Problems:

Enols, Enolates

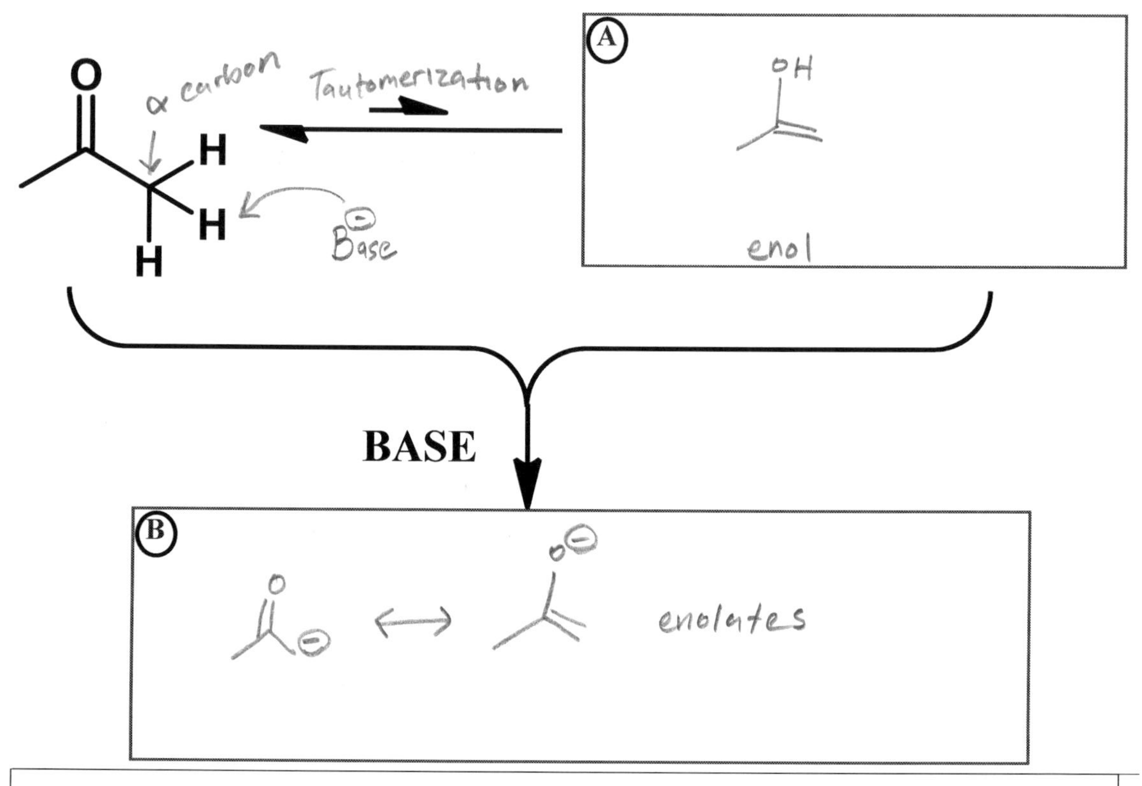

Notes:

Stabilized Enols

When evaluating the relative stabilities of species, consider all influences. The enol form can be made more favorable, for example, if the alkene in the enol is part of a π-conjugated or aromatic system:

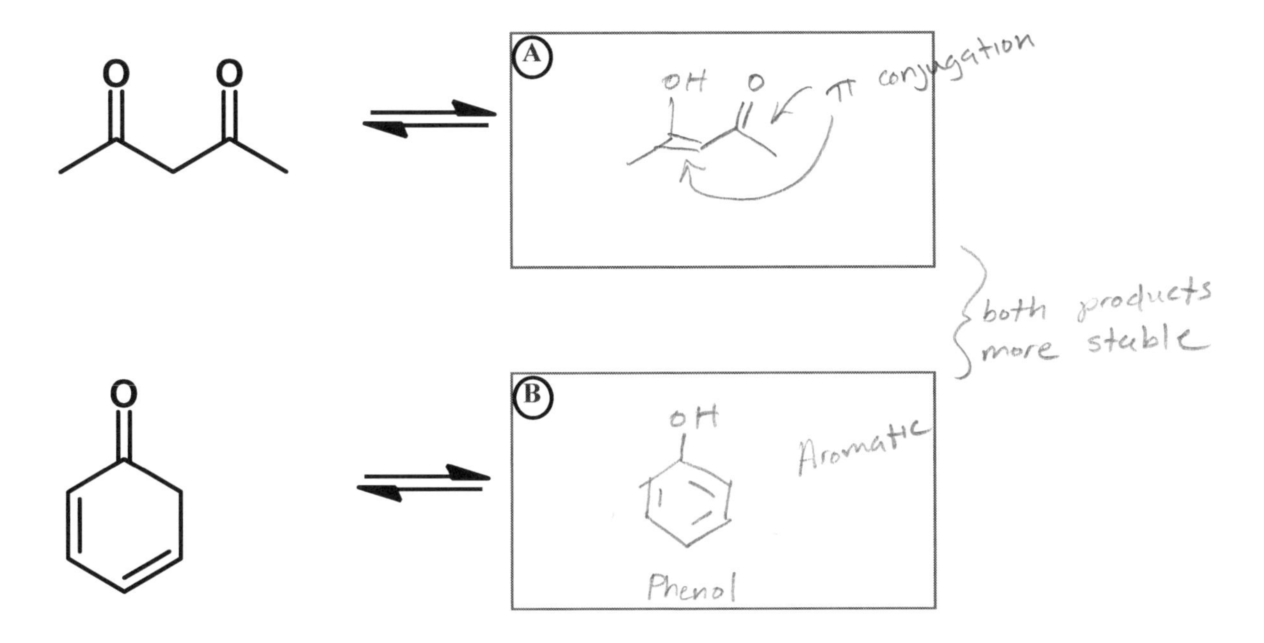

Notes:

Enolates

As with any acid-base reaction, the strength of the base we use to deprotonate the acid will influence the equilibrium constant (how much deprotonated form is present relative to the deprotonated):

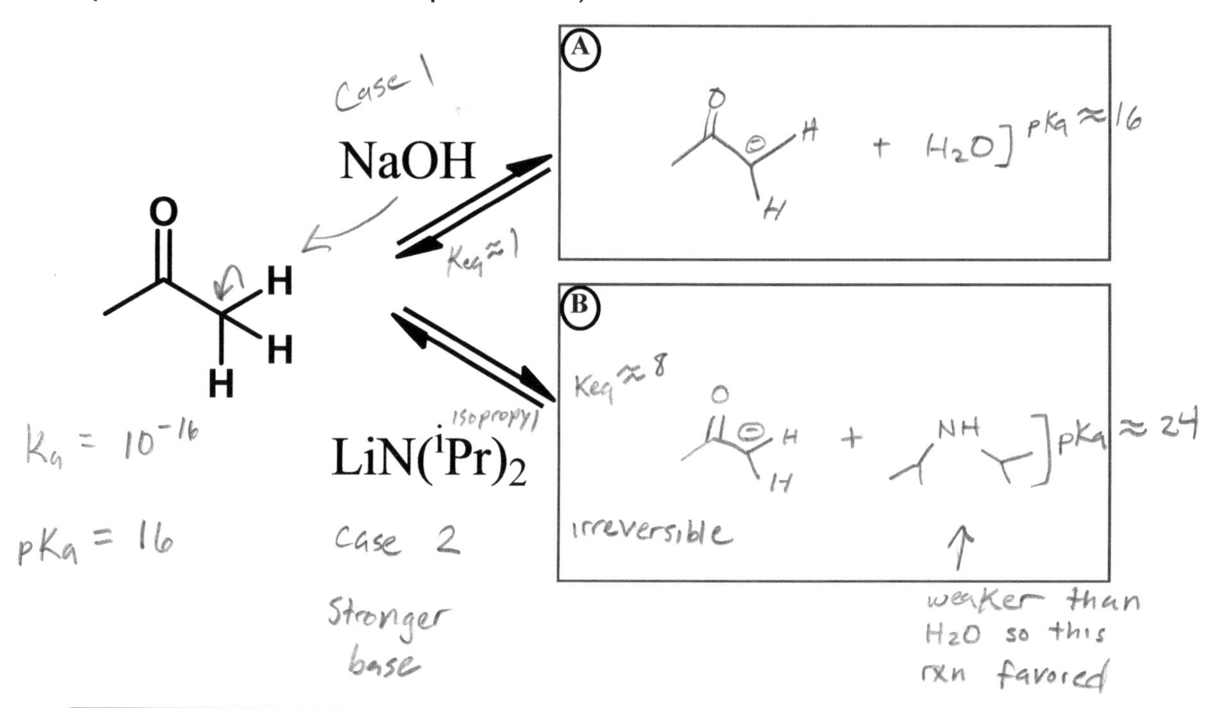

Notes:

Aldol Condensation of Aldehydes

What can we do with an enolate? One thing we can do is use it as a nucleophile to attack another carbonyl, the same way we saw for alfdehydes and ketones in the previous lecture set:

A

Δ, ^{-}OH

B

The first step is called the **Aldol Addition**; the combination of the first and second step is called an **Aldol Condensation**.

Notes:

alkene so make sure stable (when you make product)

Cont'd

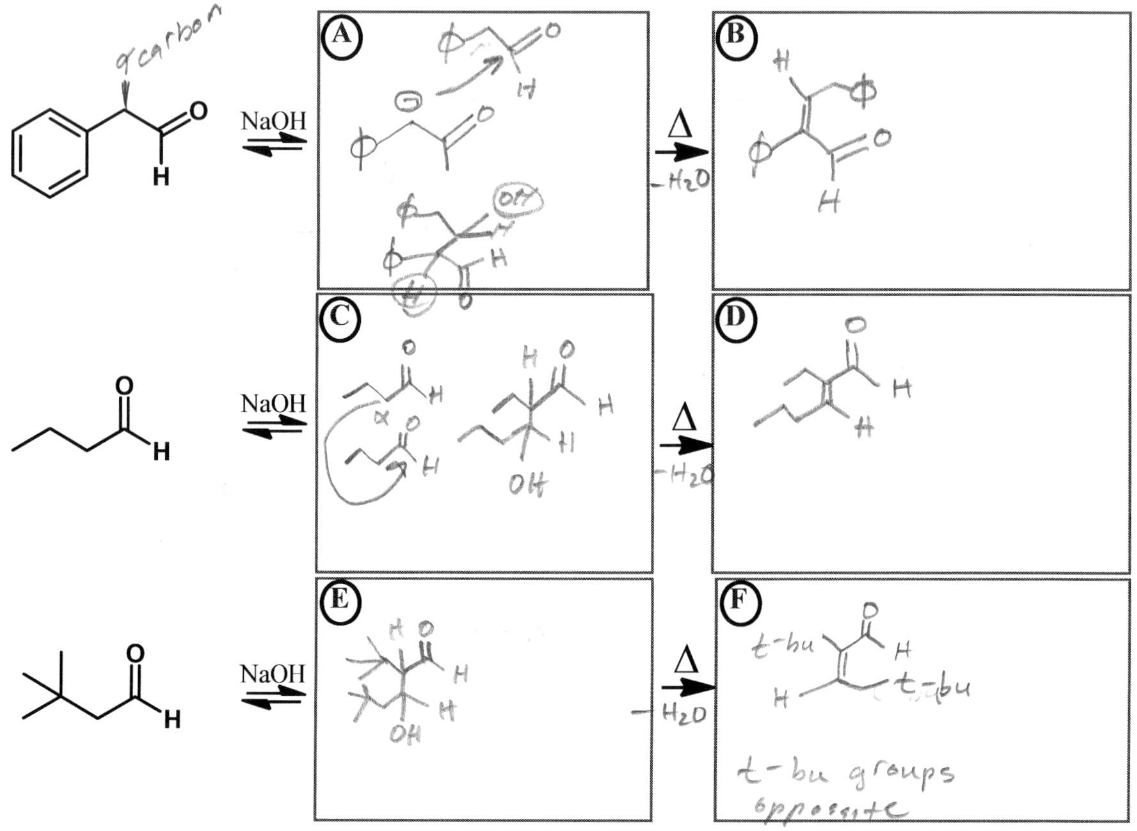

α carbon

O

H

NaOH

A

ϕ O H

ϕ O

ϕ OH H

ϕ H

H O

Δ

$-H_2O$

B

H ϕ

ϕ O

H

O

H

NaOH

C

O H

α O H

H O H

H

OH

Δ

$-H_2O$

D

O

H

H

O

H

NaOH

E

H O

H

H

OH

Δ

$-H_2O$

F

t-bu O

H

H t-bu

t-bu groups opposite

Notes:

Mechanism

NaOH

Ph

O

α

A

deprotonation

B

H_2O

C

Aldol Addition

NaOH Base Δ

D

E

Aldol Condensation

Notes:

α site of 1 attaches to carbonyl of 2nd

ALDOL CONDENSATION

Intramolecular Aldol Condensation

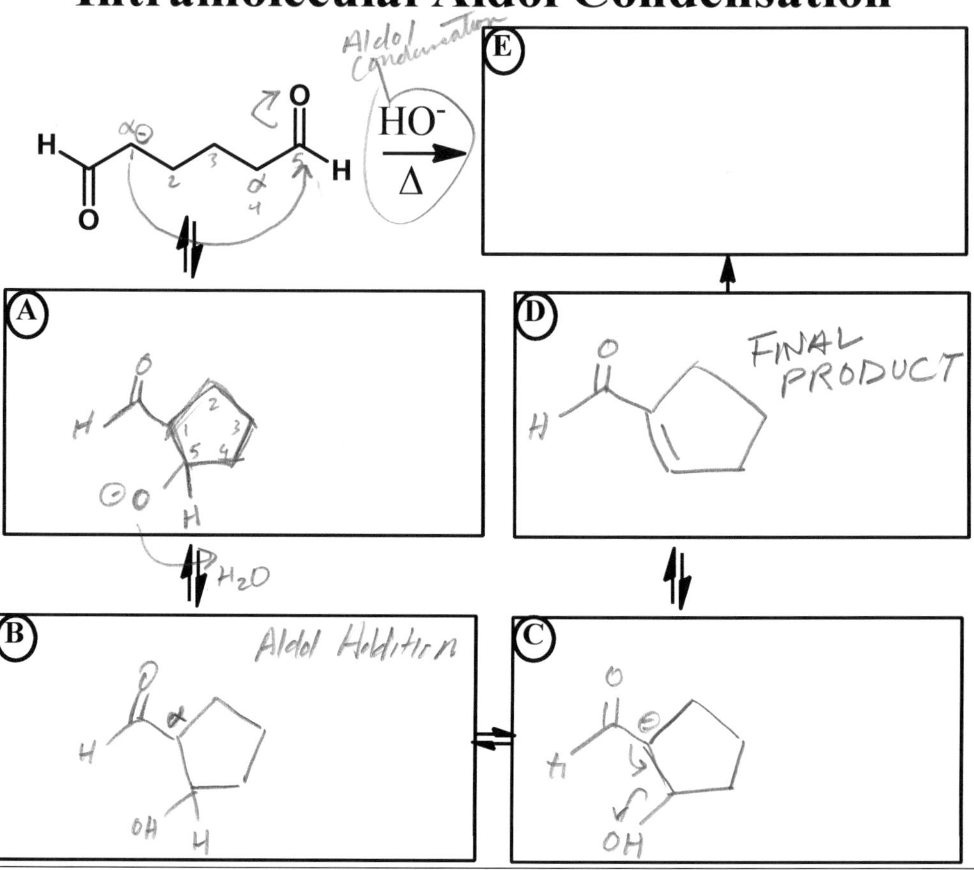

Notes:

Cont'd

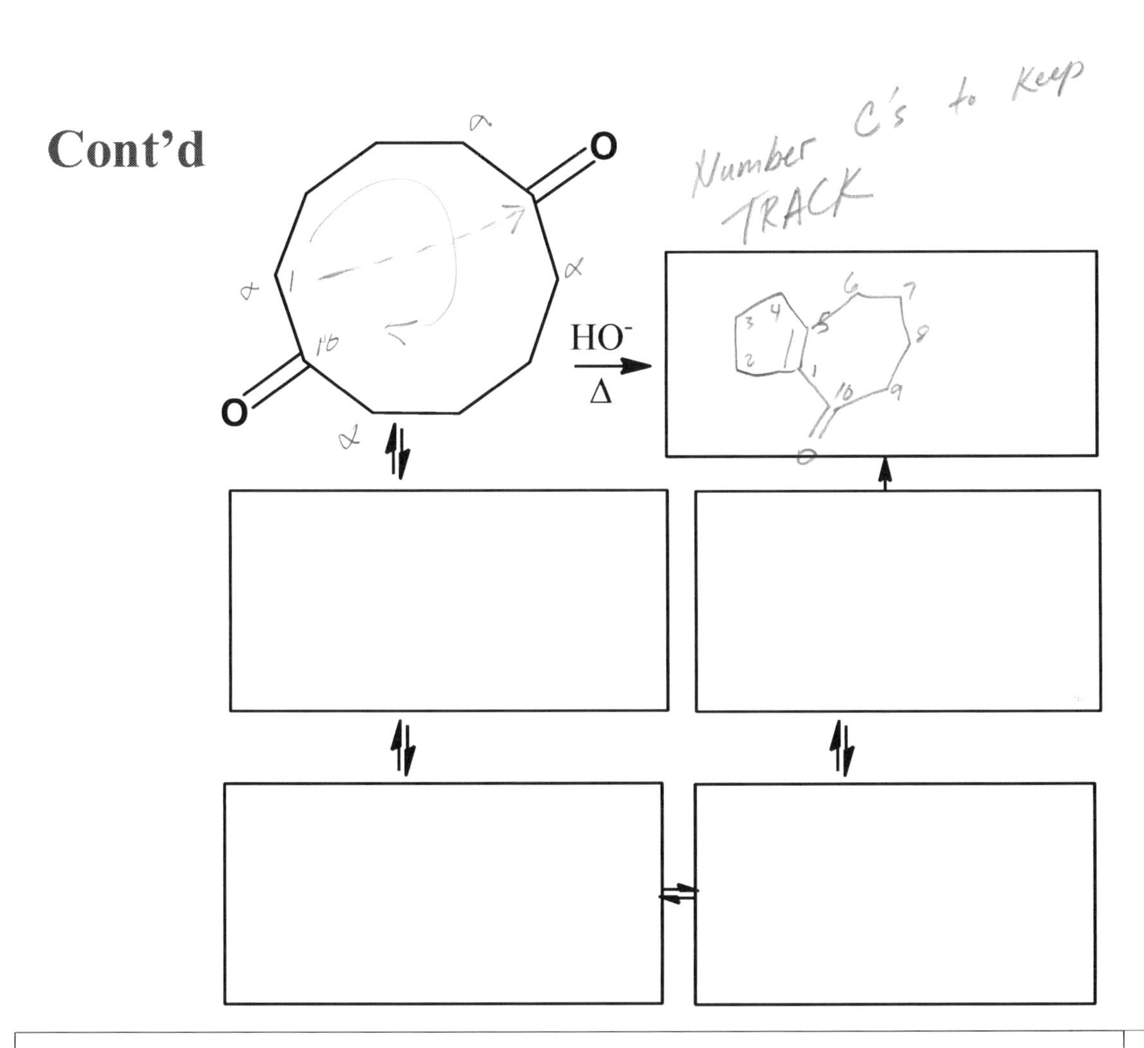

Number C's to Keep TRACK
HO⁻
Δ

Notes:

Mixed Aldol Condensation

Although the Aldol condensation is quite useful, it has its drawbacks. Consider, for example what would happen if you mixed two different carbonyls together to try to couple them:

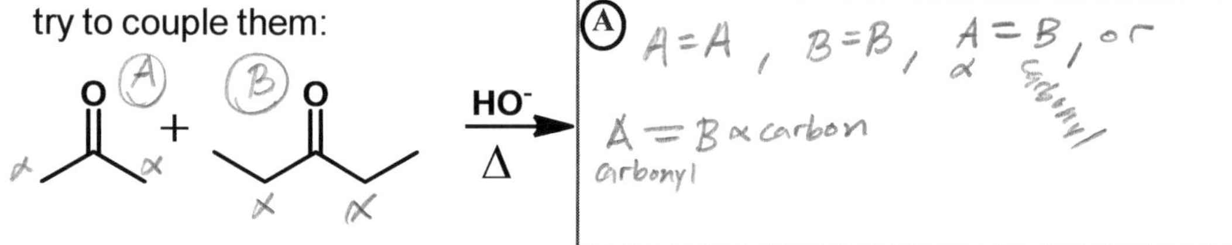

Or even one carbonyl that happens to have a different R group on each side of the carbonyl:

Notes:

Mixed Aldol Condensation

Because product mixtures result from the reactions shown on the previous page, the reaction is often cleanest when there is only **ONE type** of enolizable (alpha) proton available:

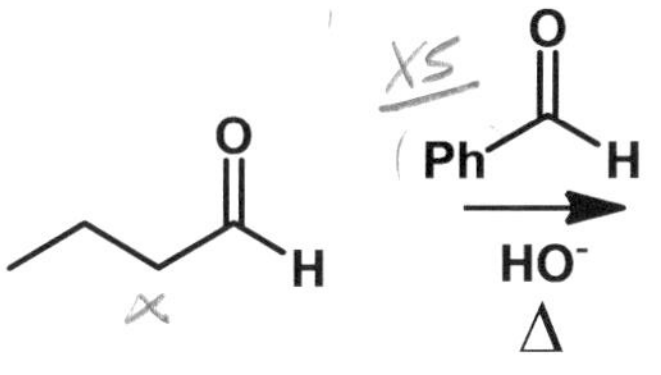

A

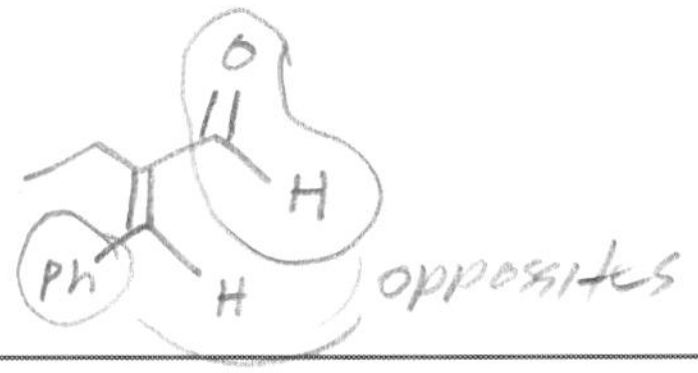

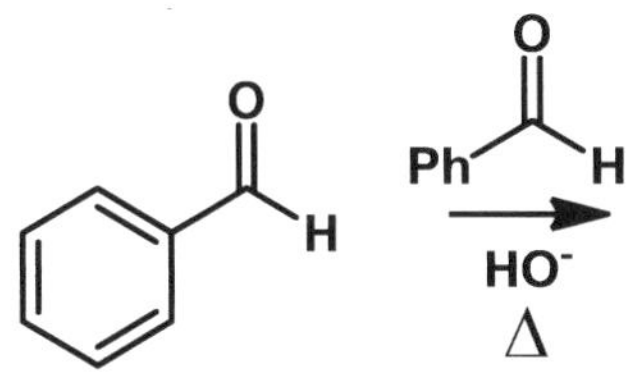

B No α H's so No Rxn

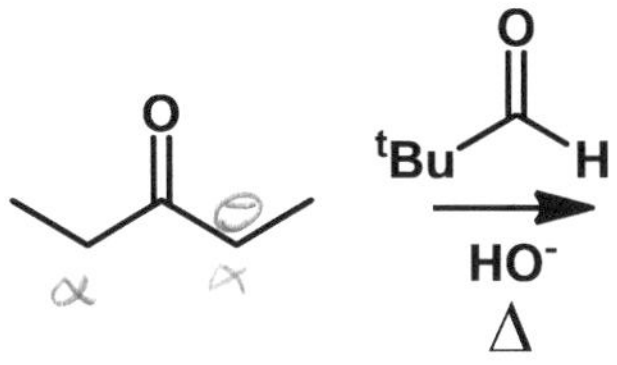

C

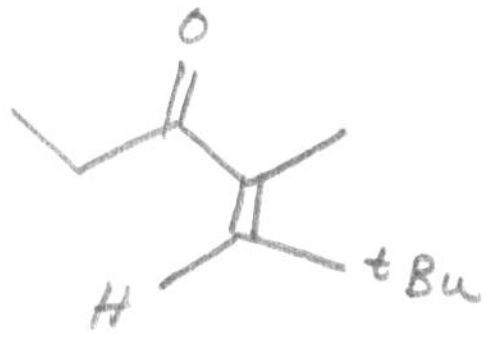

Notes:

Practice Problems

Show how these compounds could be synthesized.

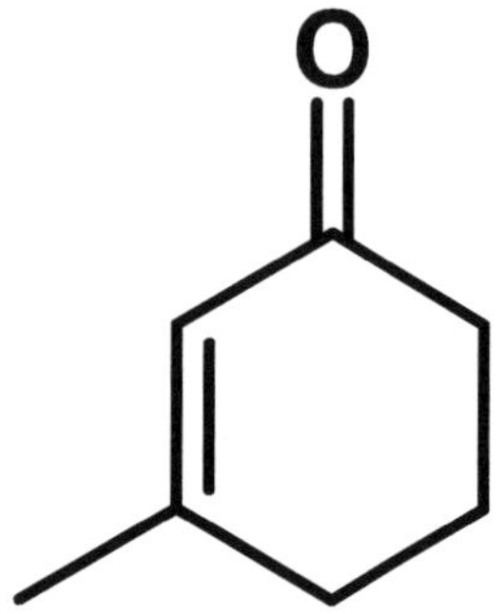

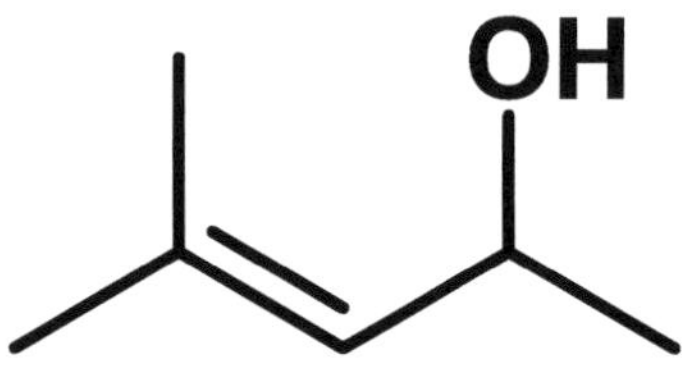

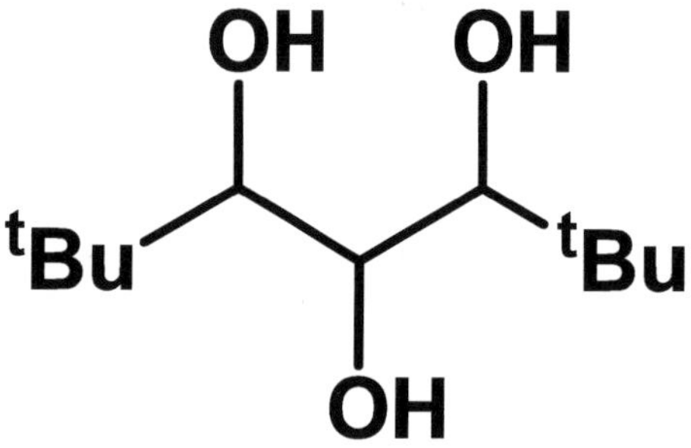

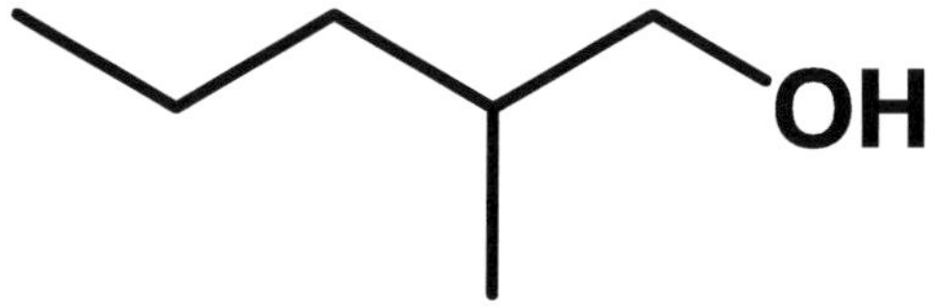

Notes:

α-Halogenation

A deprotonated carbonyl can be used as a reagent for reactions other than Aldol condensation as well. One useful reaction is α-**halogenation:**

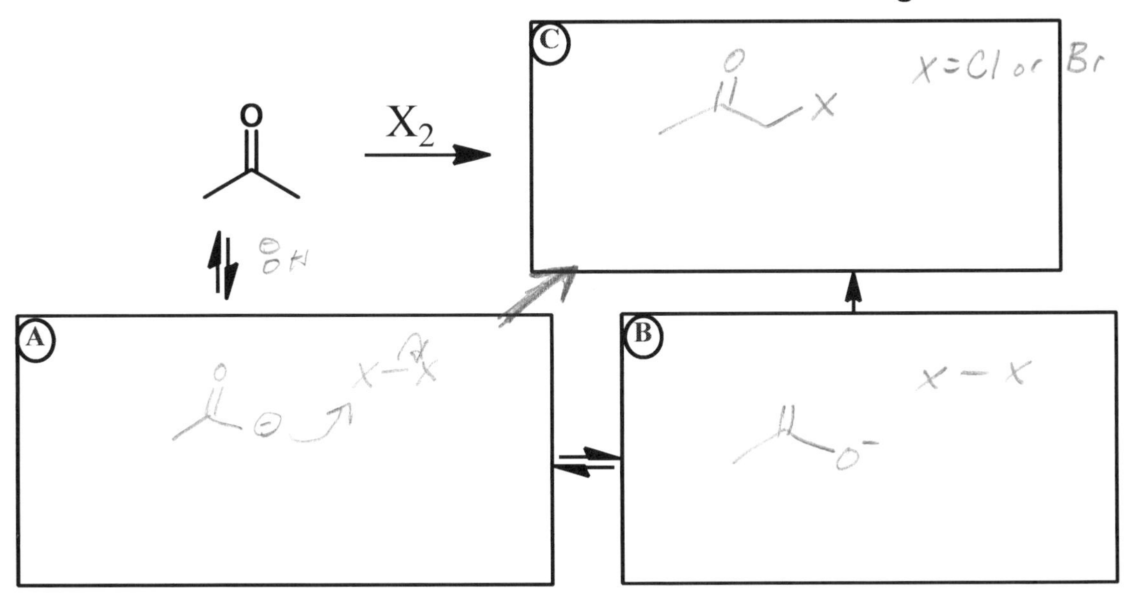

Notes:

* FINAL MATERIAL

Haloform Reaction

If we examine the mechanism in light of our previous knowledge, the origin of the products should become clearer:

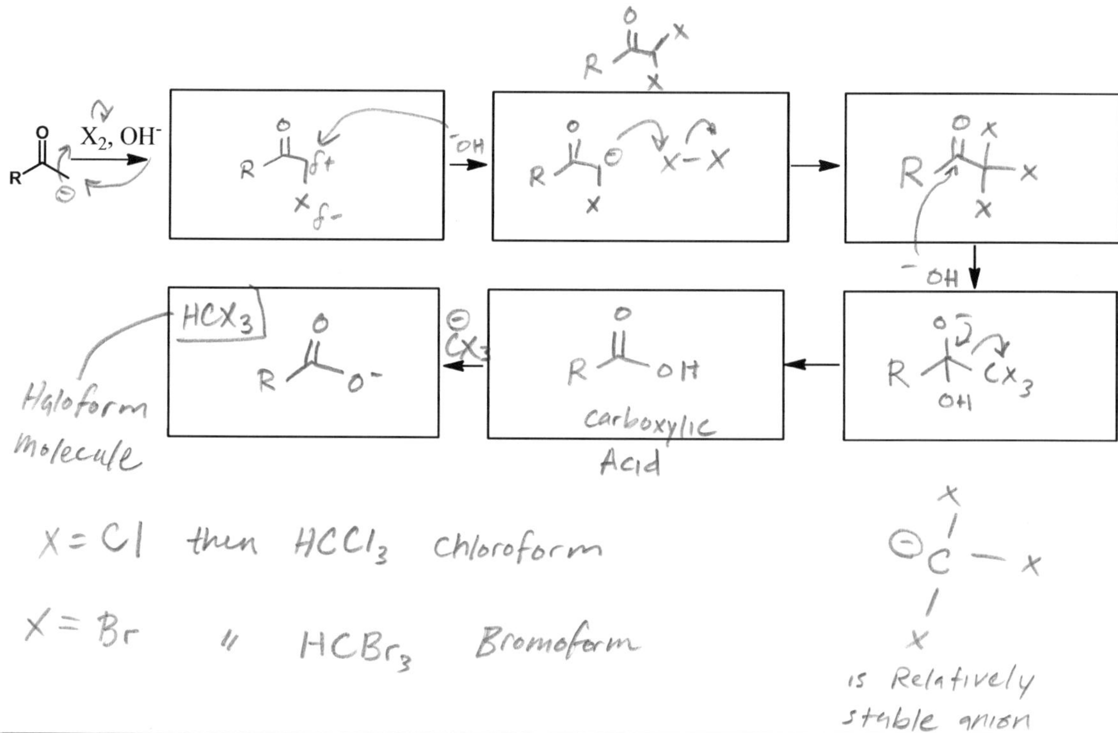

Notes:

Alkylation

Alkylation at the α position is a very useful way to add carbons to a carbony-containing molecule. The remaining carbonyls can then be reacted by other reactions we've seen to build very elaborate molecules.

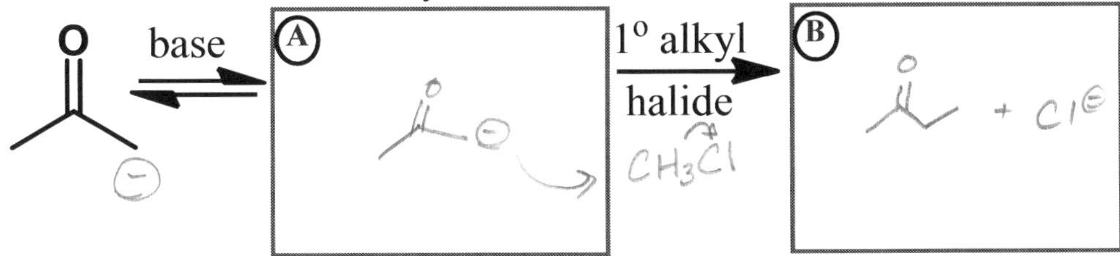

Notice, however, that the conditions are similar to Aldol condensation, which often predominates in competition. That is why β-dicarbonyls work much better for alkylation:

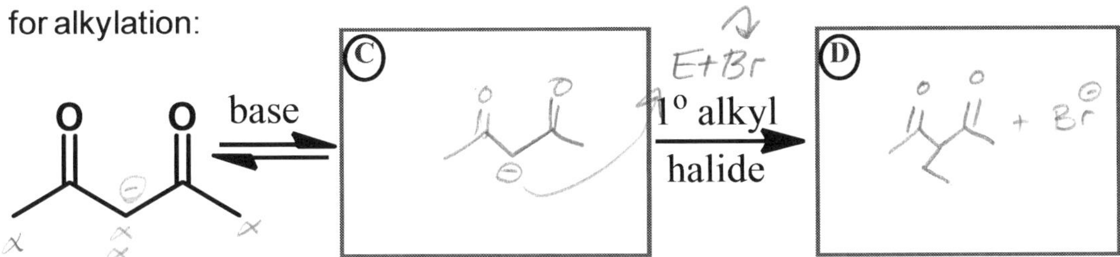

Notes:

Alkylation in Two-Step Synthesis

Here is an example where a more complex molecule is assembled from rather simple starting materials using alkylation and Michael addition in sequence:

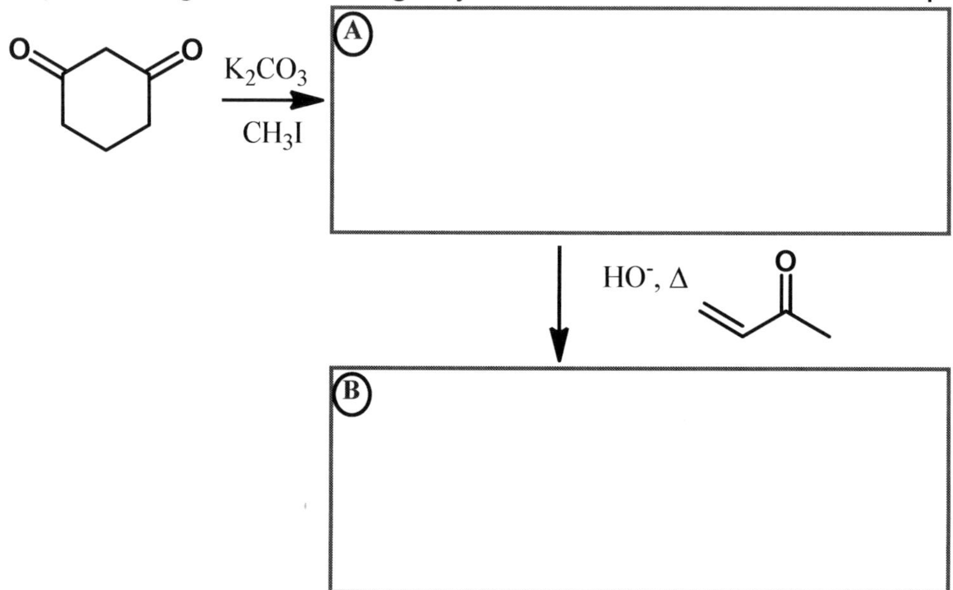

Notes:

Alkylation Pitfalls

In a simple alkylation, mixtures of products may be formed:

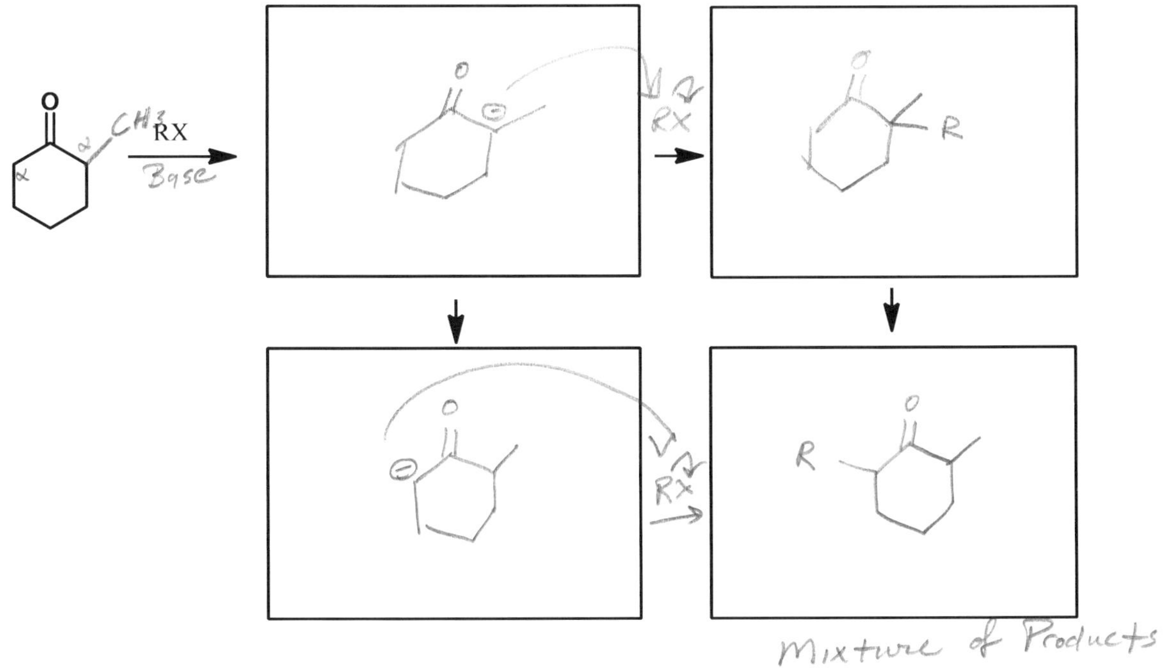

RX

Base

RX

RX

Mixture of Products

Notes:

Halogenation of Carboxylic Acid α Carbon

What if we want to halogenate the alpha position of a carboxylic acid? If we add a base to a carboxylic acid, it deprotonates the O of the COOH unit, because it is more acidic than is the alpha proton site:

(A)

So, more clever reaction conditions must be used. The specific reaction required is the **Hell-Volhard-Zelinski** (**HVZ**) reaction:

Because we already know that PBr_3 reacts with a carboxylic acid to make an acyl bromide, the addition of Br_2 must be quite important for successfully acquiring the desired alpha-bromo carboxylic acid ...

Notes:

Hell-Volhard-Zelinski Mechanism

The mechanism should make a lot of sense to us. We know that PBr_3 reacts with the carboxylic acid to make the acyl halide; we know that carbonyls with alpha-protons are in equilibrium with enol forms via tautomerization; and we know that double bonds react with Br_2:

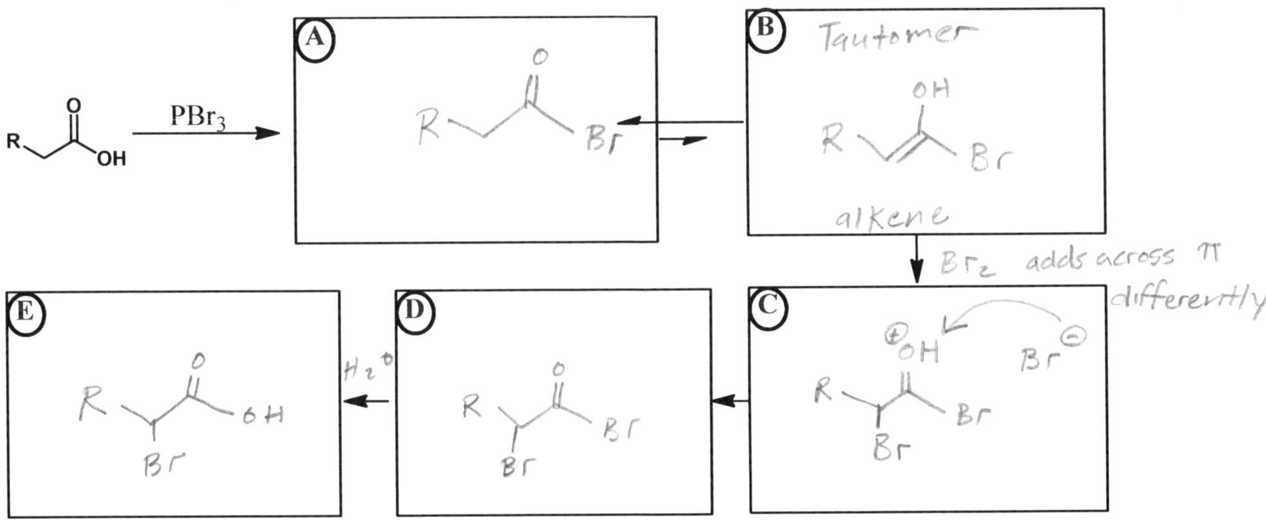

Notes:

Alkylation Pitfalls

In a simple alkylation, mixtures of products may be formed:

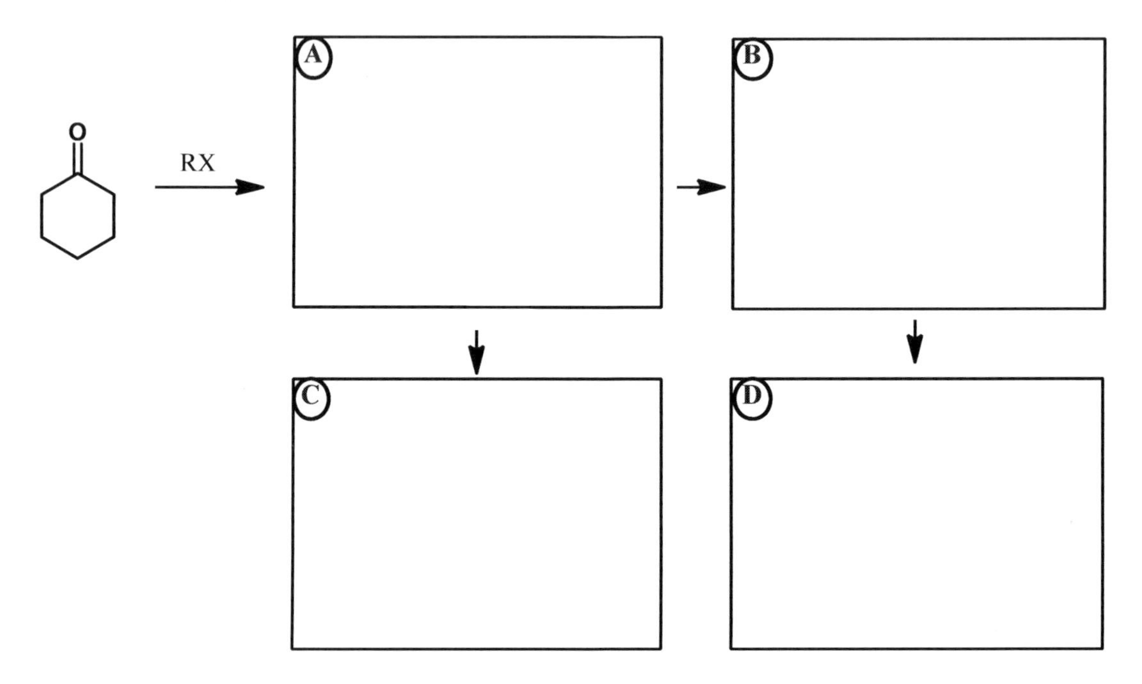

Notes:

Alkylation Fixes

A clever strategy to control substitution is to use a hydrazone in place of the carbonyl. This allows one to take advantage of **directed lithiation** from interaction of the N lone pair with Li of the base (nBuLi). The result is that monoalkylated products are obtained and, if the starting material is assymetric, preference for substitution at the less substituted side:

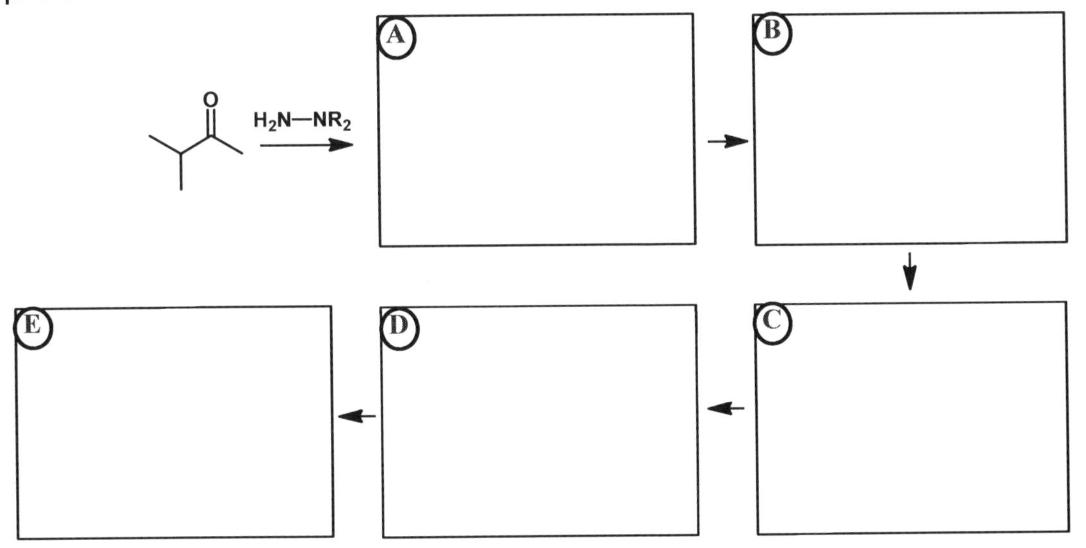

Notes:

Alkylation Fixes

Another strategy **to obtain monoalkylated** product is to **use an enamine** as your nucleophile in place of the enolate. Because the enamine (like the hydrazone) is easily prepared from the carbonyl, this is a convenient protocol:

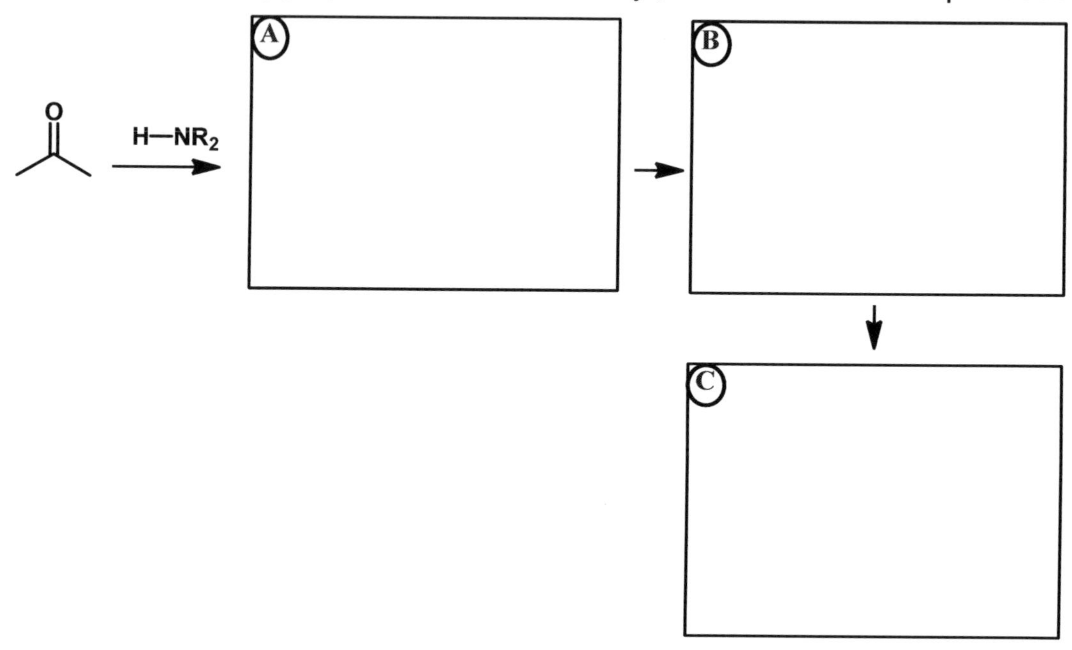

Notes:

Lecture Set 9:
Reactions of α,β-Unsaturated Carbonyls

Suggested Reading:

Suggested Problems:

α,β-Unsaturated Aldehydes and Ketones

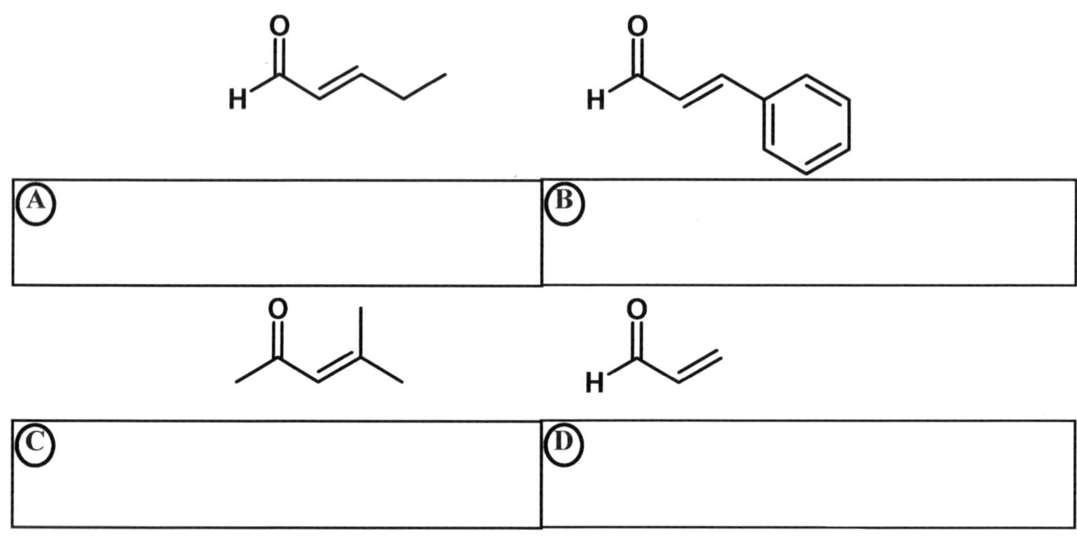

When we look at α,β-unsaturated carbonyls, we must consider the effect that the pi-conjugation of the C=C and C=O bond and the potential resonance structures of intermediates might have on the reactivity of this class of compounds.

Notes:

Preparation of α,β-Unsaturated Carbonyls

We've already seen how some α,β-unsaturated carbonyls can be made via the **Aldol condensation**

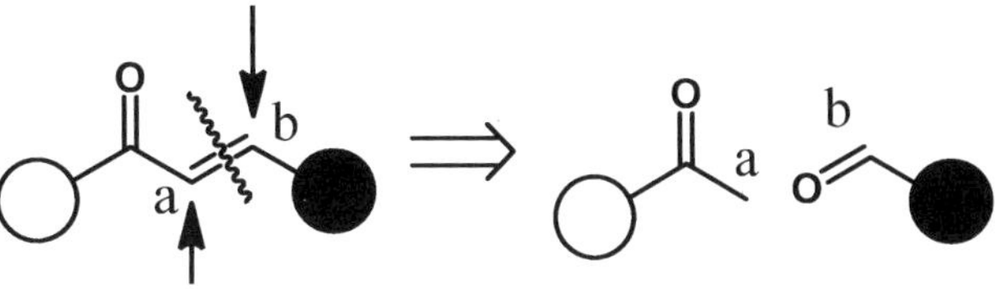

H, O ⟹ (A)

O ⟹ (B)

H, O ⟹ (C)

H, O ⟹ (D)

Notes:

Conjugate Addition

We have seen enough examples to know how carbonyls react with nucleophiles and how an alkene might react with a nucleophile:

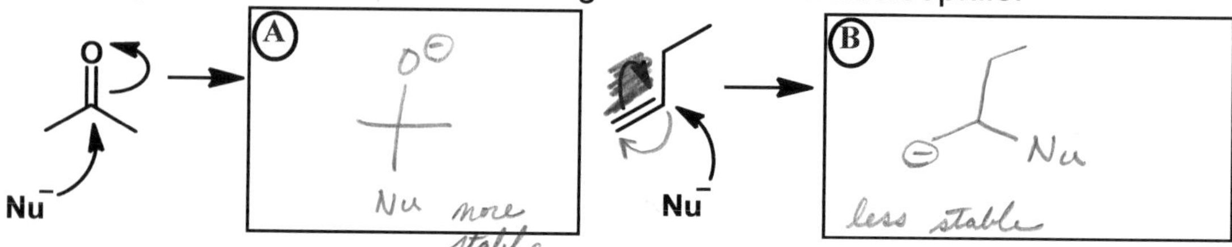

If the carbonyl and the alkene are in pi-conjugation with one another, let's see what the impact might be:

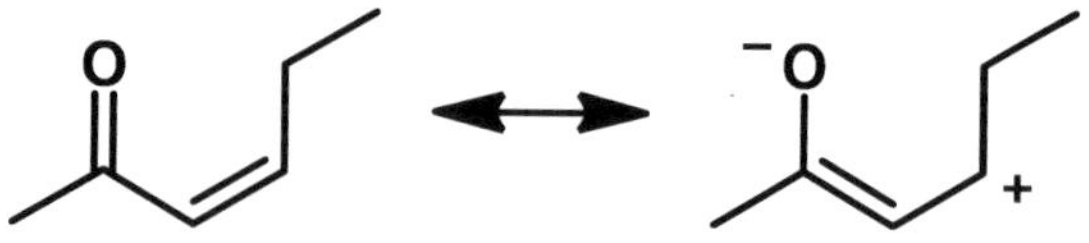

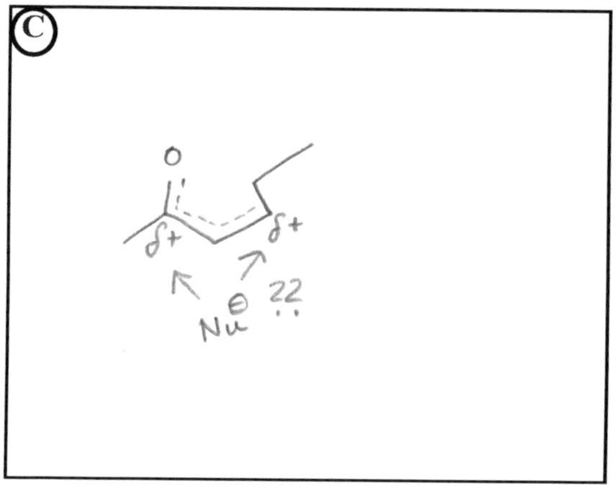

Notes:

α,β-Unsat. : Conjugate addition

As a result of the **two electrophilic sites** present in an α,β-unsaturated carbonyl, there are two potentially competing pathways for reaction with a nucleophile:

1,2-addition (direct addition)

A

Nu⊖ → Nu, O⊖, H⁺ → OH, Nu

1,4-addition (direct addition)

B

O, Nu⊖ → Nu, OH, enol ⇌ Nu, O, H

Tautomerization

Notes:

Strong Bases Give Direct Addition

Fortunately, the type of addition depends in large extent on whether the attacking nucleophile is a strong or weak base:

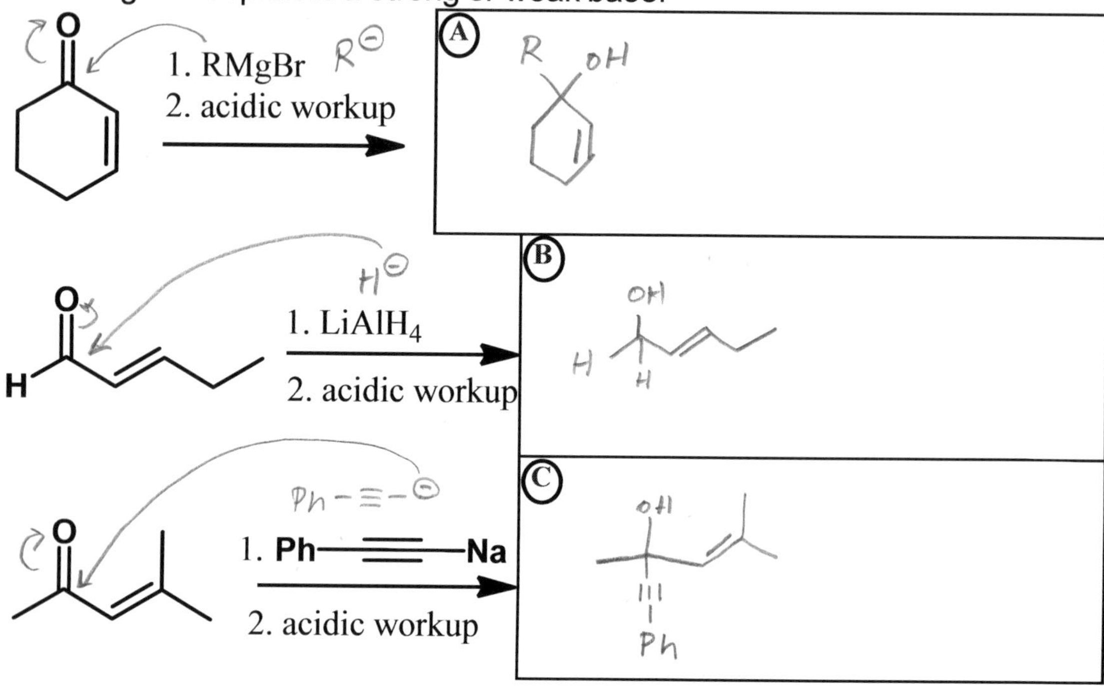

Notes:

Weaker Bases tend to give Conjugate Addition

It's a good thing we know our acid-base rules!

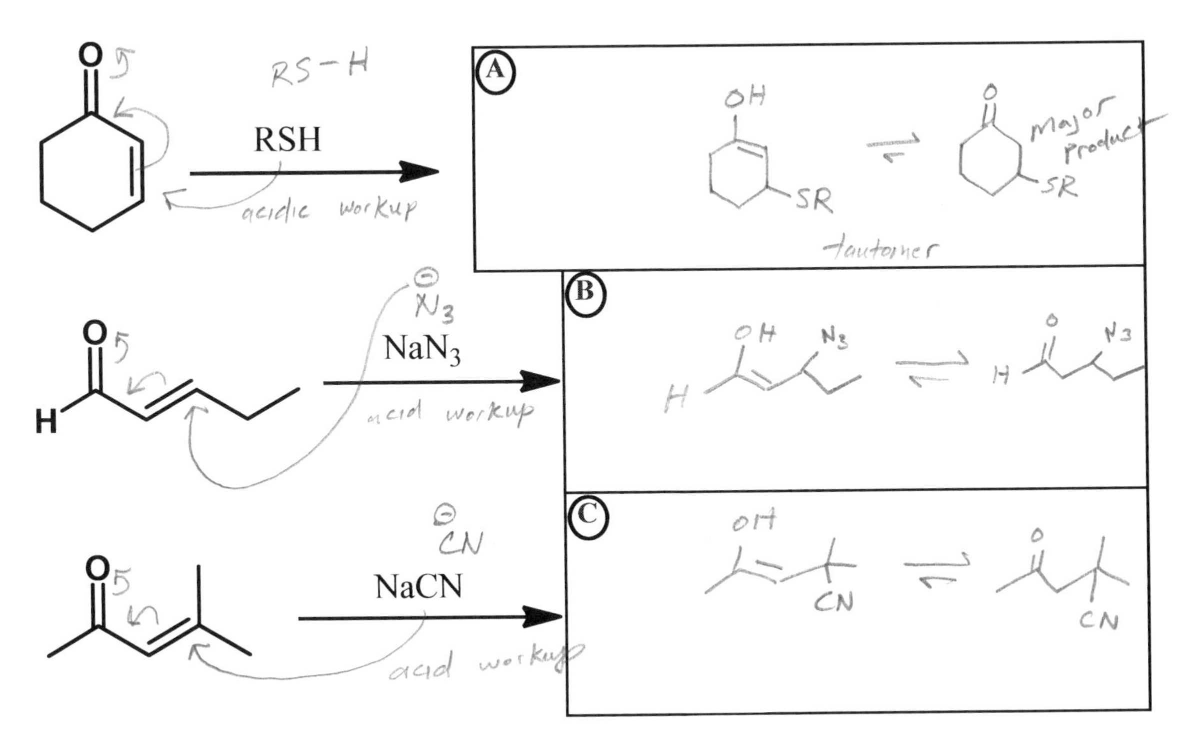

Notes:

Michael Reaction: Enolate for conjugate addition

Let's use our newfound knowledge of how unsaturated carbonyls tend to react to predict what might happen if an unsaturated carbonyl reacts with an enolate. First, let's evaluate the enolate's relative base strength:

(A) enolate ≈ $^{\ominus}OH$ < $^{\ominus}NR_2$ < $R^{\ominus}$

weak base ... Strong Base

Now we should be able to predict what an enolate nucleophile would tend to do upon reaction with an α,β-unsaturated carbonyl:

R ... R' ... (B) ... R ... R' ... O ... $H^{\oplus}$

(should have one more carbon in chain)

OH ... R ... R' ... O

Taut.

O ... R ... R' ... O

Notes:

Michael Reaction: Enolate for conjugate addition

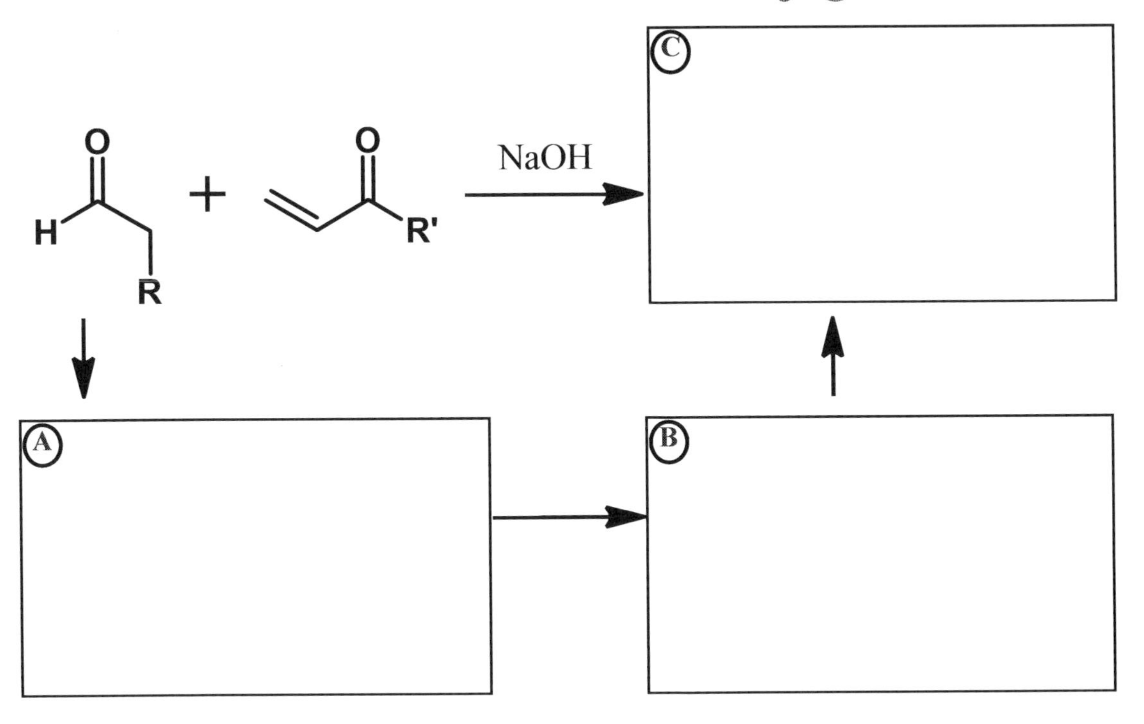

Notes:

Michael Reaction: Enolate for conjugate addition

You may notice that the conditions we use for the Michael reaction (strong base) are the same type of conditions used for the Aldol condensation. For this reason, the reactions can be in competition with each other. The Michael reaction can be made more favorable if we use a starting material that is more completely converted to the enolate form. For this reason, one commonly uses β-diketones as the enolate component of a Michael addition:

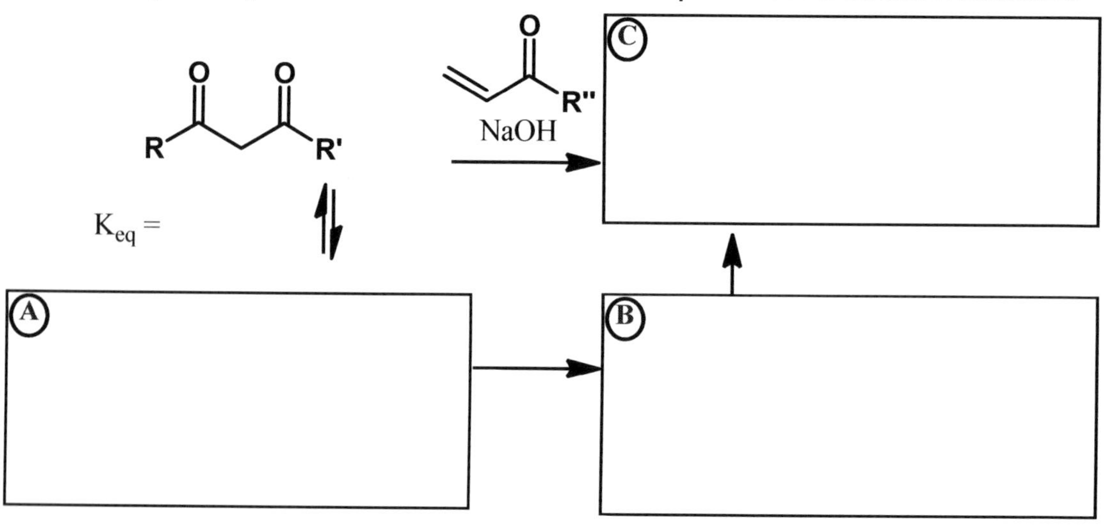

Notes:

Michael then Aldol = Robinson

You may notice that the product of a Michael reaction is a species that contains at least two carbonyls in it. This means that we can do additional reactions between the carbonyls on the initial Michael product. If one does a Michael addition followed by an Aldol condensation, this is the **Robinson Annulation**, an important route for making substituted cyclohexanones:

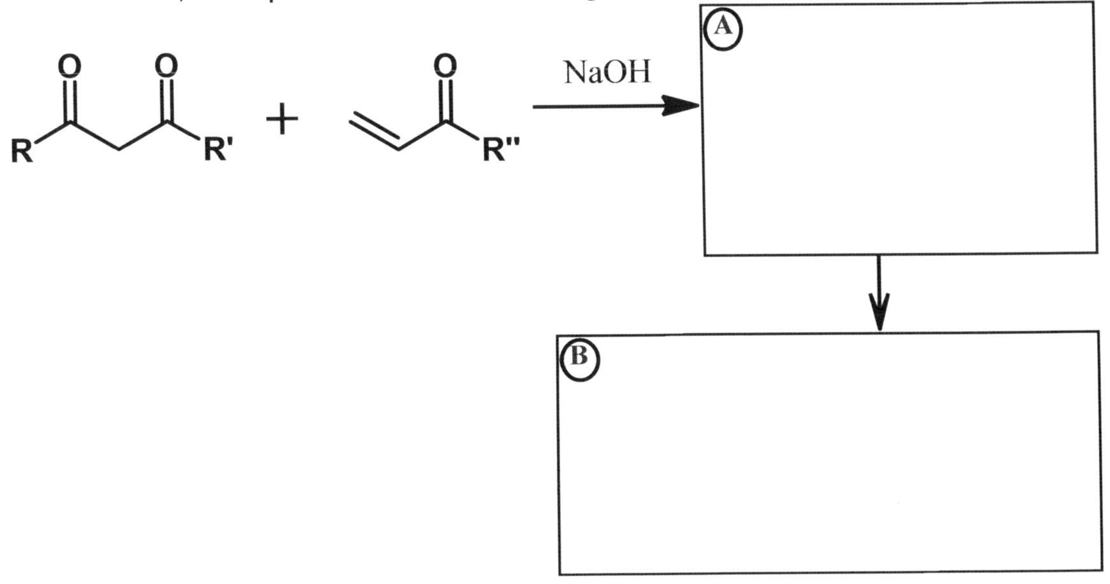

Notes:

Michael then Aldol = Robinson

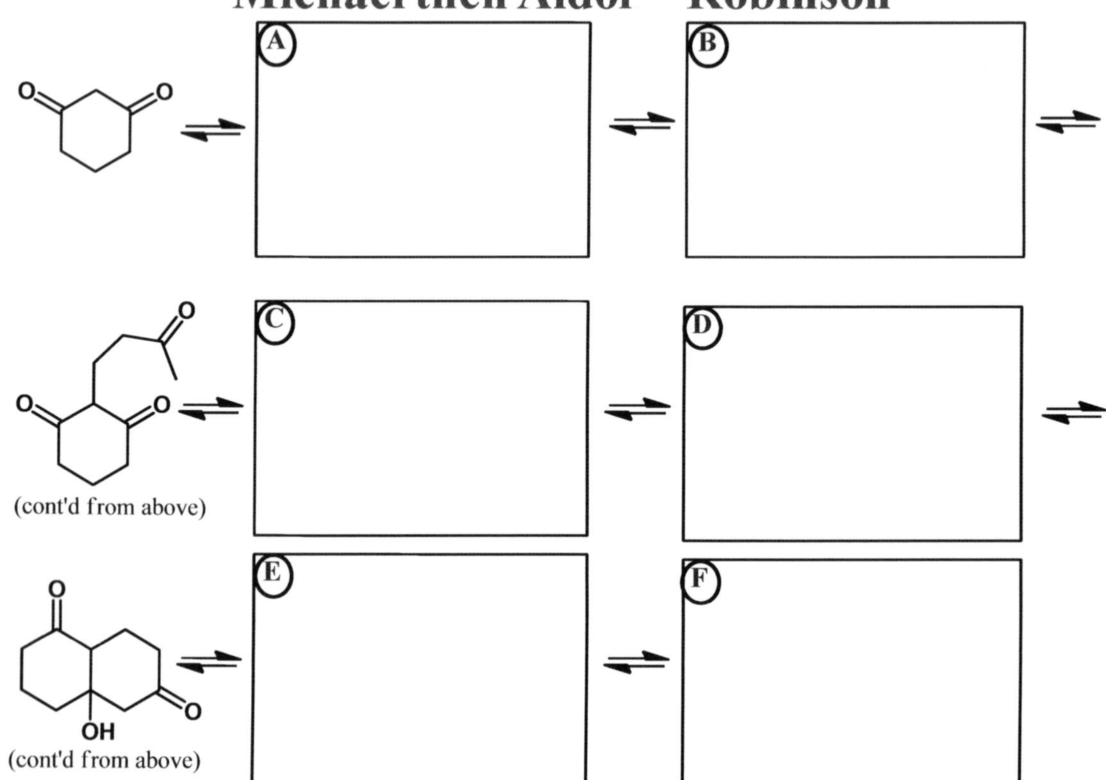

Notes:

Practice

Using reactions covered in this lecture set, propose a viable route to prepare each of these materials:

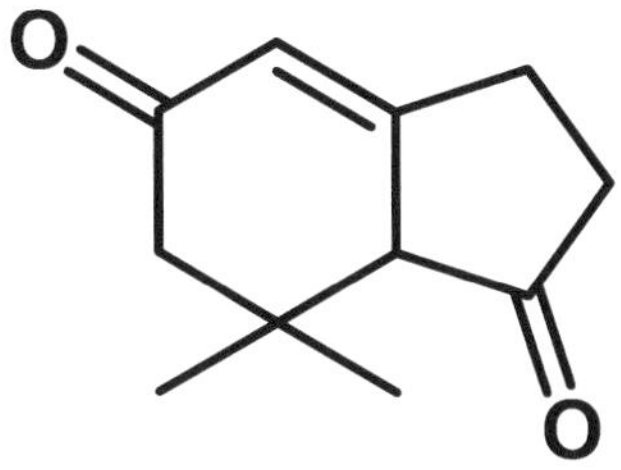

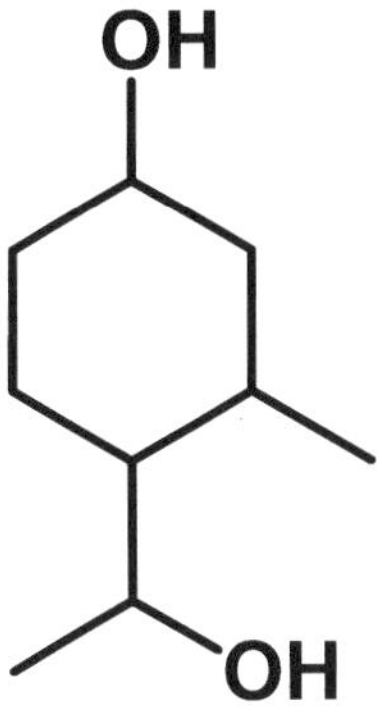

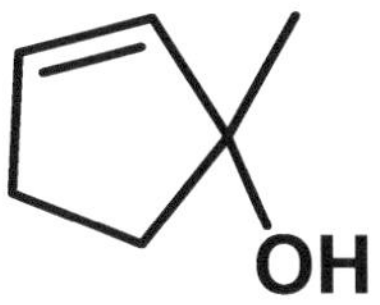

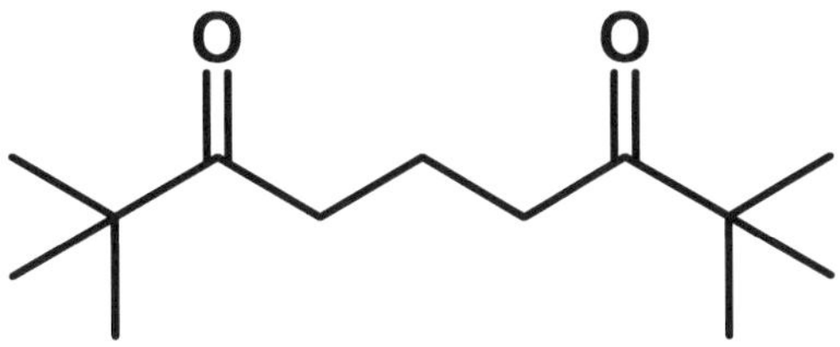

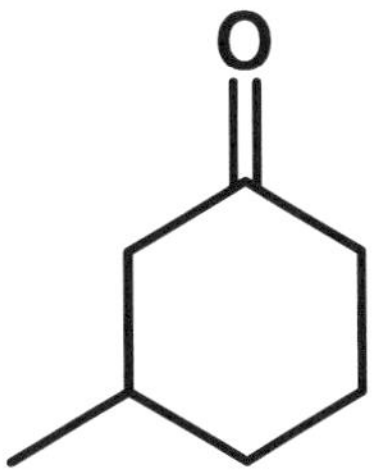

OH

O

O

OH

O

O

O

OH

Notes:

Stork Enamine Reaction

Another variation on the Michael reaction is the **Stork Enamine Reaction**. This is exactly like the Michael reaction, but it starts with an enamine in place of the enolate (we have seen this strategy before for alkylation). Because a base is not used, the Aldol condensation is not a competitive process:

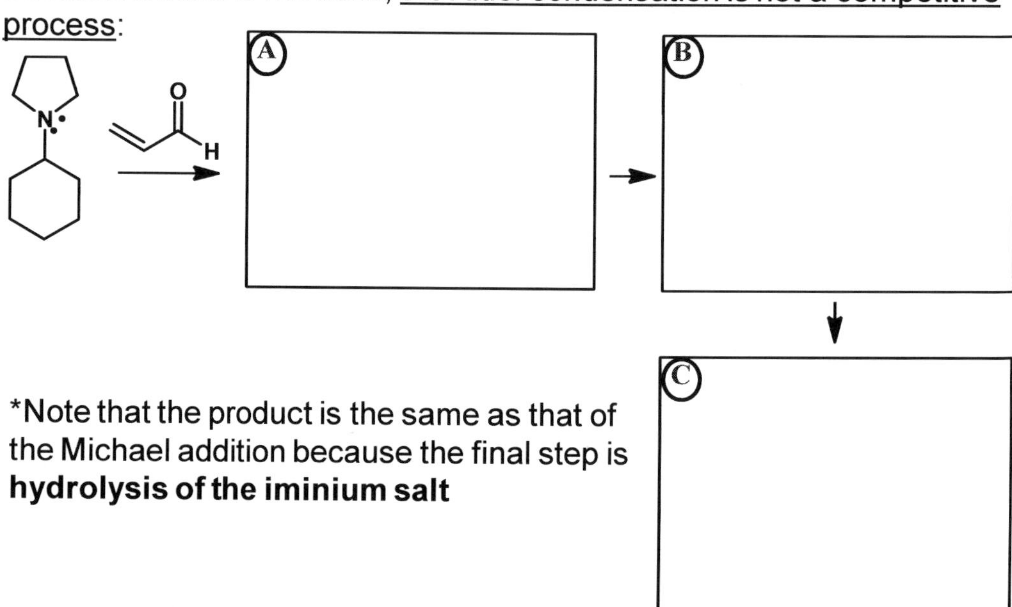

*Note that the product is the same as that of the Michael addition because the final step is **hydrolysis of the iminium salt**

Notes:

Claisen Condensation

We have seen that in the Aldol condensation, the first step is a simple addition of the enolate nucleophile to the ketone or aldehyde just like we saw for other nucleophiles adding to aldehydes or ketones. Now, we are going to look at using an enolate to add to an ester. This is called the **Claisen Condensation**. What would we predict would happen using what we already know?

2 (ester, OR) → 1. RO^- 2. H^+ → A → B → C

Still in Basic media

Aldol Rxn Addition

**Notice that we're using a stronger base than HO^-

Notes:

Some General Trends in Carbonyl Reactivity

Recall our two general fates for carbonyls once the nucleophile has attacked:

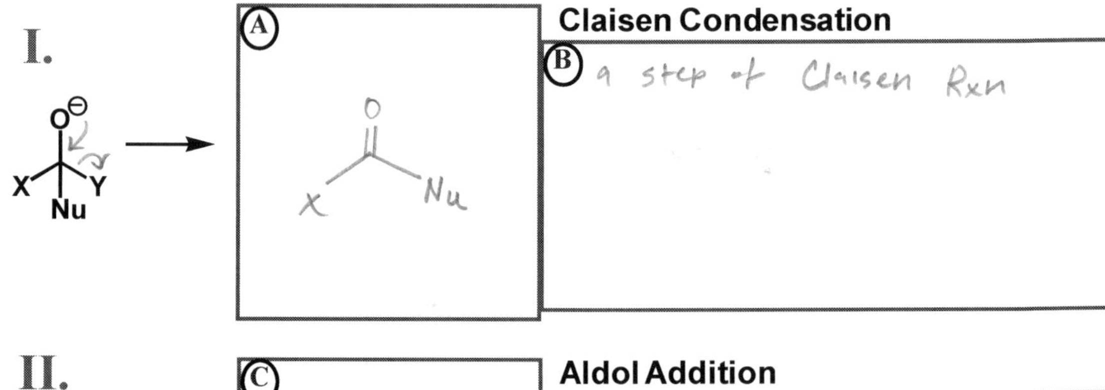

Notes:

Mixed Claisen Condensation

Just as we saw with the Aldol condensation, the Claisen condensation can be used to couple two different esters. Also analogous to the Aldol condensation, the mixed Claisen condensation works best when:

(A) only one ester has α-proton then mixed Claisen Rxn is easily done

O O
OR + Ph OR
1. RO^-
2.H^+

(B) Ph OR → Ph OR

Think about why one would want to add the first reagent slowly…

Notes:

Almost a Claisen Condensation

A reaction that is essentially the same as the Claisen reaction involves using an enolate from a ketone or aldehyde as the nucleophile to attack an ester's carbonyl carbon:

works w/ other good LG?

also what about direct Nu addition

1. HO^-
2. H^+

remember that "R" can be some ugly HC

A

Ph

Note that the difference in **anion stability** of the enolate derived from a ketone or aldehyde versus that derived from an ester leads to a strong preference for deprotonation at the alpha carbon of the ketone or aldehyde:

B

Notes:

Intramolecular Claisen = Dieckmann

If a claisen condensation happens to occur between two esters that are on the same molecule, this is called a **Dieckmann condensation** (for whatever reason):

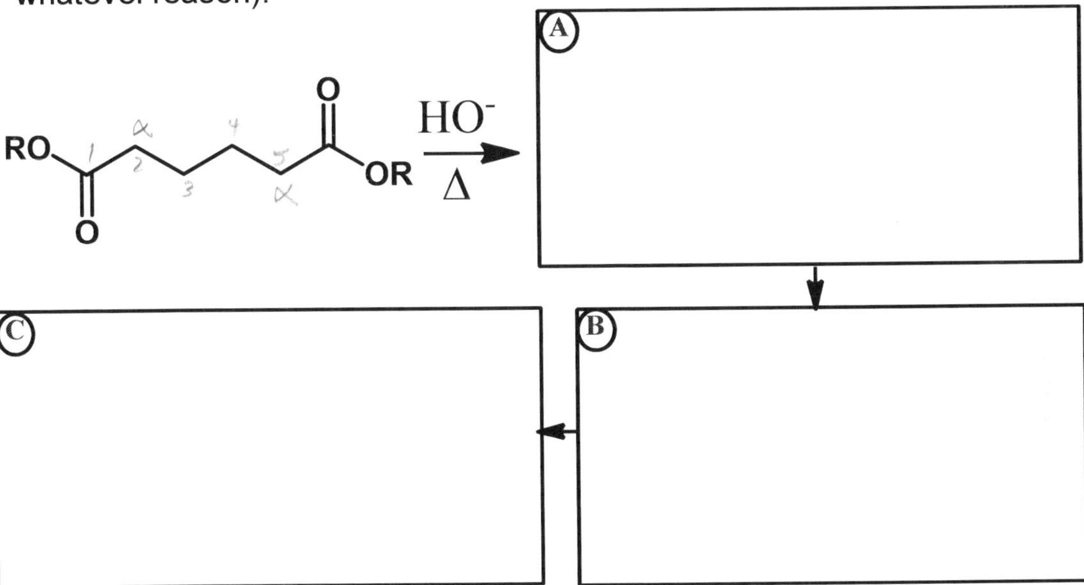

Notes:

Decarboxylation

We have seen many reactions in which carbonyl compounds are amended with additional groups or transformed into other functional groups. Sometimes we would like to use this versatility to make a complex molecule, but would then like to remove the carbonyl. This can be done by a process called **decarboxylation**. This is often done simply by heating. The temperature required varies on the basis of the functional group. The 3-oxo-carboxylic acids are relatively easy to decarboxylate:

Not Resonance arrow

A Δ CO_2 + enolate — $H^{\oplus}$ taut. — B acetone + CO_2

The addition of catalytic acid can facilitate the process via intramolecular proton transfer:

OH $H^{\oplus}$ Δ C + CO_2 — taut. — D + CO_2

Notes:

Decarboxylation Application I: Malonic Ester Synthesis

Now that we know that a 3-oxo carbonyl is often relatively easy to decarboxylate, we can exploit this property in some specific and very useful reactions. The first of these that we will cover is the **malonic ester synthesis**. This reaction typically starts with diethyl malonate and is used to prepare various carboxylic acids:

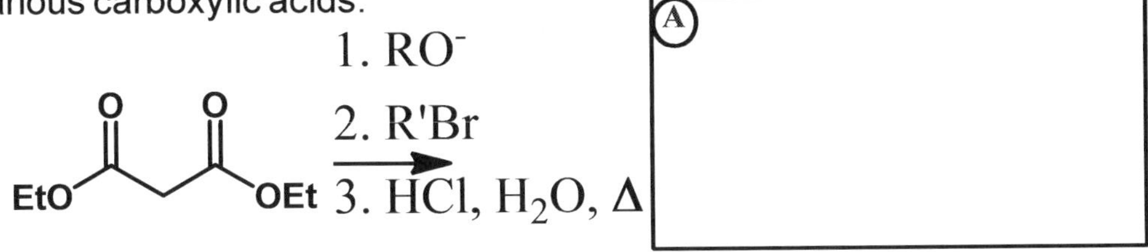

We should be able to reason through what's happening in these three steps.

Step 1. Base:

Step 2: R'Br:

Step 3. Acid and heat:

a.

b.

Notes:

Mechanism for Malonic Ester Synthesis

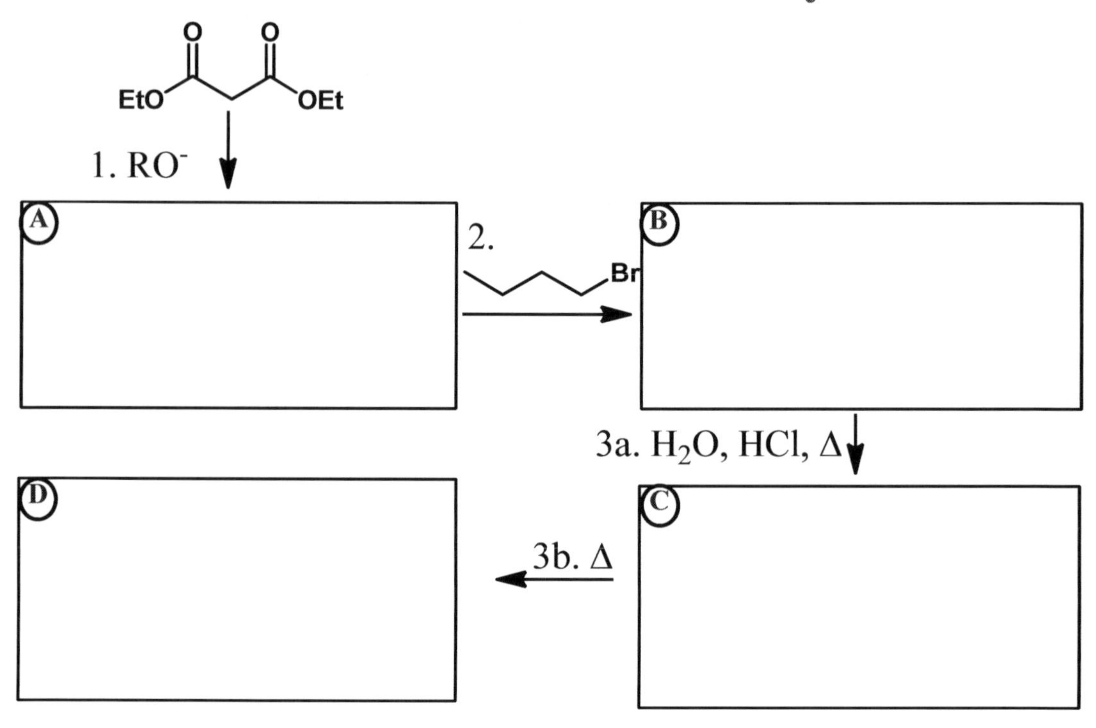

Notes:

Decarboxylation Application II: Acetoacetic Ester Synthesis

Now that we have seen the malonic ester synthesis, the **acetoacetic ester synthesis** should be relatively straightforward. This reaction simply has a methyl group in place of one of the two ethoxy groups of diethylmalonate:

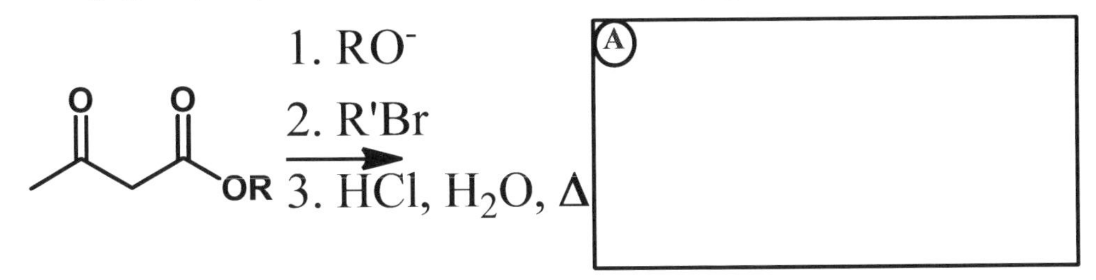

The three steps are quite analogous to those we saw for malonic ester synthesis as well…

Step 1. Base:

Step 2: R'Br:

Step 3. Acid and heat:

a.

b.

Notes:

HUNTER 100 — 10pm

~~Monday 4/18~~ — $Nu^{\ominus}$ addition to Carbonyls

~~Tuesday 4/19~~ — " sub. " "

~~Wednesday 4/20~~ — Enol / Enolates Rxns

~~Thursday 4/21~~ — Exam 3

Tuesday 4/26 — Additional Topics

Wednesday 4/27 — Mock Exam

Thursday 4/28 — Work for Mock Exam

Monday 5/2 — Final Exam Hints

5/4 — FINAL EXAM

5 Pts. EC for each session

Made in the USA
Lexington, KY
16 December 2010